SOUND KNOWLEDGE

SOUND KNOWLEDGE

Music and Science in London,
1789–1851

Edited by
JAMES Q. DAVIES
and ELLEN LOCKHART

THE UNIVERSITY OF CHICAGO PRESS
CHICAGO AND LONDON

The University of Chicago Press, Chicago 60637
The University of Chicago Press, Ltd., London
© 2016 by The University of Chicago
Published 2016.
Printed in the United States of America

25 24 23 22 21 20 19 18 17 16   1 2 3 4 5

ISBN-13: 978-0-226-40207-9 (cloth)
ISBN-13: 978-0-226-40210-9 (e-book)
DOI: 10.7208/chicago/9780226402109.001.0001

Library of Congress Cataloging-in-Publication Data
Names: Davies, J. Q., 1973– editor. | Lockhart, Ellen, editor.
Title: Sound knowledge : music and science in London, 1789–1851 /
    edited by James Q. Davies and Ellen Lockhart.
Description: Chicago ; London : University of Chicago Press, 2016. |
    © 2016 | Includes index.
Identifiers: LCCN 2016009803| ISBN 9780226402079 (cloth : alkaline paper) |
    ISBN 9780226402109 (e-book)
Subjects: LCSH: Music and science—History—18th century. | Music and
    science—History—19th century. | Music—England—London—18th century—
    History and criticism. | Music—England—London—19th century—History
    and criticism. | Science—England—London—History—18th century. |
    Science—England—London—History—19th century.
Classification: LCC ML3805 .S656 2016 | DDC 780/.05—dc23
    LC record available at http://lccn.loc.gov/2016009803

♾ This paper meets the requirements of ANSI/NISO Z39.48-1992
(Permanence of Paper).

CONTENTS

Introduction: Fantasies of Total Description   *1*
JAMES Q. DAVIES AND ELLEN LOCKHART

1.   Music as an Object of Natural History   *27*
   EMILY I. DOLAN

2.   Celestial Mechanisms: Adam Walker's Eidouranion,
   Celestina, and the Advancement of Knowledge   *47*
   DEIRDRE LOUGHRIDGE

3.   Transparent Music and Sound-Light Analogy ca. 1800   *77*
   ELLEN LOCKHART

4.   Charles Wheatstone: Musical Instrument Making,
   Natural Philosophy, and Acoustics in Early Nineteenth-
   Century London   *101*
   MYLES W. JACKSON

5.   Charles Wheatstone's Enchanted Lyre and the Spectacle
   of Sound   *125*
   MELISSA DICKSON

6.   Instruments of Empire   *145*
   JAMES Q. DAVIES

7.   Good Vibrations: *Frankenstein* on the London Stage   *175*
   SARAH HIBBERD

8.   Engine Noise and Artificial Intelligence:
   Babbage's London   *203*
   GAVIN WILLIAMS

9.   Hearing Things: Musical Objects at the 1851 Great Exhibition   227
            FLORA WILLSON

        Acknowledgments   247
        Contributors   249
        Index   251

Fantasies of Total Description

JAMES Q. DAVIES AND ELLEN LOCKHART

"The forming of the five senses," Karl Marx famously theorized in 1844, "is a labor of the entire history of the world down to the present." A frequent visitor to London in these years, Marx imagined that sensuous capacities themselves were radically transformed by shifting modes of capitalist production. One of the principal fantasies of the *Economic and Philosophic Manuscripts* was that physiology itself was subject to vast historical movements in industrial property relations. This meant claiming that an emerging "slave class" had been overtaken by the mere "sense of possessing, of having." "The sense for music," "a musical ear," or "an eye for beauty of form," he argued, could be forged or withheld in relation to dark bourgeois forces. Such was the depravity of the midcentury experience, apparently, that "to the eye an object comes to be other than it is to the ear, and the object of the eye is another object than the object of the ear."[1] For Marx, this emergent autonomy of the senses was by no means to be celebrated. Their estrangement one from the other, instead, betrayed the alienation of these "slaves" from their "authentic" physical and relational natures.

The Marxist avowal of prelapsarian sensory life is pure wish fulfillment, as much as the fallacy of sensory reconfiguration or realignment. How exactly do the senses undergo metamorphosis? What exactly does it mean to speak of "an anthropology" or "a history" of the senses? The "modern" isolation of aural perception in the period of Marx's writing will be familiar to music scholars: from tales of the general emergence of concert-hall listening, or yarns about sonorous artworks emerging as fully autonomous objects of disciplinary knowledge. Marx's cogitations might be taken to suggest that the evolution of such purely sonic matters of concern be narrated in relation to grand bio-cultural shifts in the very structure of human hearing. He believed—as many have since—that everything felt and known could be ex-

1

plained in relation to deep transformations in economic, technological, and media conditions. It was only in full view of sensory history, Marx might say, that the enormous expansion of the sense for sound in the period around 1844 could be explained, as much as the prevailing British imperative (discussed in what follows) not merely to "have" music, but to have it all. His was always a narrative of implausibly epic proportions.

The present volume, which addresses music and science in London during the first half of the "long" nineteenth century, worries about such totalizing fantasies. Our purview is framed by two attempts to "have it all"—that is, two attempts at total description. We begin with Charles Burney's monumentally ambitious *General History of Music*, a four-volume study of music around the globe "from the earliest ages to the present period," published between 1776 and 1789. And we conclude with "the world" coming to London for the Great Exhibition of 1851, to witness the accoutrements of "all music"—alongside other instruments of science and industry—pinioned in the immense glass-encased collections of the Crystal Palace. Marx's narrative of progressive sensory alienation sheds gloomy light on these monuments—as it might, if invited to do so, on the English invention of "the aural" (a category pertaining exclusively to "the organ of hearing") in the late 1840s, and on many other trophies from this period's extraordinary acceleration of musical activity and scientific innovation. The grandiosity of these weighty fantasies left its traces in our language in ways too easily overlooked or taken for granted. In addition to "the aural," the period we are considering here bequeathed us the "scientist" (1834), as well as "musicology" (1845, from the German *Musikwissenschaft*) as a term and as a discipline.[2] It also saw the emergence of *energy* in what is now its most familiar sense, as a term to denote quantities of work power. It gave us the "stock exchange"; the discipline of acoustics, led by "the acoustician"; and an early glimpse of the microphone. It saw the emergence of the electric telegraph as the first means of rapid long-distance communication. If the native musical talent of this nineteenth-century metropolis bequeathed no works to the current canon—a fact that did much to earn England the title of *das Land ohne Musik*— its robust institutions of music can be credited with decisively fostering the very canon itself, as well as commissioning significant works by composers from afar. The baton conductor appeared in London during these years, as did the proms concert. The list could go on. We are not concerned here, however, with providing a comprehensive account of the landmarks of this extraordinary period; rather we ask how sonic and musical practices, performers and performances, instruments and technologies structured emergent disciplinary knowledge and emergent disciplinary divisions.

A central motivation for this book was the need to rebalance the voluminous academic literature on visual culture (in studies of popular science and spectacle in London) in favor of "aural culture."[3] The idea was to refocus scholarly attention, in the hope that musicologists, with their training in aesthetic practice and histories of listening and performance, had insights to offer historians of science and technology, particularly given mutual interests in questions of sensory experience. How can science be listened for and known aurally? How might "auditory culture" or the study of musical practice be usefully brought to bear on research into historical ways of seeing and knowing? The facts and figures gathered in these pages thus tend toward accounts of embodied knowledge. They grapple with physical experience first and address the instrumental conditions that make epistemological inquiry emotionally compelling and technically possible.

A central aim of this volume is to locate "London," in its fullest phenomenological sense, as a city formed in relation to sites at once musical and scientific. Of course, the metropolis could be mapped in numerous ways. One could chart the rise of training institutions and pedagogical infrastructures, or the rapid expansion of a global market for technologies alongside burgeoning markets for commercial retail, scientific, and music publishing in advertising, periodical literature, newsprint, and educational literature. One could document the flood of new instruments and communications media, considering systems such as the kaleidoscope, thaumatrope, phenakistoscope, photograph, difference engine, telegraph, typewriter, and stereograph, alongside other enchanting communication technologies—the lyrichord, coelestinette, celestina, eidophusikon, harmonium, concertina, and myriad more or less exotic musical instruments. Or, alternately, our "London" might be understood as multiple networks of amateur and professional musicians and scientists, interweaving in that city's uniquely versatile spaces; this approach would involve tracing the interactions between music makers, natural historians, astronomers, music historians, engineers, explorers, missionaries, composers, publishers, acousticians, instrument manufacturers, and entrepreneurs. What London was, in other words, cannot be thought as separate from its mode of becoming, the "modern" disciplinary, technical, and epistemological accumulations that characterize our chosen period.

The findings of this volume show that concern for music and concern for science were often one and the same; the differences between "optical" and "auditory" inquiry, between "music" and "science," between what counted as "musical performance" and what counted as "scientific performance" were often difficult to define. The activities of the characters that emerge as protagonists of our chapters attest to this: people such as Charles Burney, Adam Walker, Thomas Young, Nicola Sampieri, Charles Babbage, and Charles Wheatstone enjoyed multifaceted careers as entrepreneurs and lecturers, inventors and musicians, performers and instrument manufacturers, teachers and travelers, performers and publishers. The range of their public and private endeavors betrays the extent of interaction between domains of inquiry, and between the metropolis and the provinces.

The concern to situate metropolitan knowledge in relation to specific sites of experience responds to various currents in scholarship.[4] Take, for instance, increasing interest in the geographies of both music and science—of exhibition and concert spaces as overlapping domains. Note, too, resonances between, on the one hand, a growing body of literature on the spectacular exhibitions of popular science, and (on the other) an equally energetic musicological literature on virtuosity and spectatorship.[5] Part of the function of this volume, then, is to register a disciplinary convergence. It is also, pace Marx, to show that visual and aural cultures worked in tandem rather than independently. The audiovisuality of both scientific and musical practice, as has become clear, means that aural and visual cultures were hardly as distinct as later assumptions would suggest, and that all these purportedly distinct sensory "cultures" mattered together.

Where did music and science meet? A dizzying variety of arenas existed in London for the purposes of scientific and musical activity, sites that played host to mobile and ever-shifting audiences as practitioners vied at once to entertain and enlighten. The very concept of science was bewilderingly heterogeneous, as any consideration of its assorted public, educational, and commercial venues would suggest. There were competing claims to authority, the terms of natural law being fought over by utilitarians, natural philosophers, establishment Tory-Anglicans, Oxbridge academics, "gentlemen of science," dissenters, evolutionists, evangelicals, radicals, and numerous other nonconformists. The institutions of science, the metropolitan calendar of miscellaneous lectures and *conversazioni*, and the various theaters, clubs, colleges, universities, gardens, laboratories, homes, parlors, schools, journals, workshops, and societies made for a diffuse, if not chaotic, scene.

Particularly in the early decades of our period, what counted or not as art, science, and music was neither effectively policed nor professionally maintained. The "natural philosophy" lectures of traveling luminaries such as John Arden and Adam Walker—a principal means of conveying popular science to the masses before 1800—often included a demonstration of the instruments of music and science together, and musical performances in the intervals. But music resonated even in London's temples of high science during these years, thanks in no small part to the unusual abundance of polymaths and musical amateurs among its scientific celebrities. In 1799, for instance, Thomas Young—recently advertised to the readers of popular science in the present era of big data as "the last man who knew everything"—lectured at the Royal Society on the science of organ pipes, speaking trumpets, the resonance of chords, and the correct tuning of intervals. Young's comprehensive course on natural philosophy and mechanical arts for the Royal Institution (founded in 1799 for the purpose of "applying science to the needs of the nation, and especially agriculture") included lectures devoted to sound, harmonics, musical instruments, the theory of music, and its history.[6] As early as 1803, the astronomer, pyrotechnician, and musician William Crotch played his idiosyncratic musical "specimens" while illustrating "Music considered an Art and a Science" at the Royal Institution. Later, the Royal Institution played host to the talents of Charles Wheatstone, the son of a London flute and bassoon maker, who features in three chapters at the center of this volume; his groundbreaking acoustical demonstrations at the institution relied on strange chimerical noisemakers—scientific instruments built from bastardized musical ones—as well as a Javanese *gendèr*, and an artificial voice box.

One could turn a profit from the demonstration of novel instruments in London's culture of entertainments: a fact Wheatstone knew from experience, as he had begun his career as the curator of "Charley Wheatstone's Clever Tricks," a veritable museum of self-playing instruments on display in the Great Room at Spring Gardens. As this volume makes clear, even the most purposively musical of these newly invented instruments invited attention not just to their sounds but also the way, as aural technologies, they appeared to the eyes. What was more, music featured prominently in most of the much-studied visual entertainments of this period—the myriad dioramas, panoramas, udoramas, and cosmoramas built for the interior display of landscapes, as well as automata and waxwork figures, and other feats of engineering. If one attended Madame Tussaud's exhibition in the evenings—wrote one commentator in an 1838 *Pocket Guide to London*—one could enjoy concerts performed amidst such figures as Napoleon, Joan of Arc, Frederick the Great, Henry VIII, and even Marie Antoinette, "full-length . . . dressed as in life," eerily still and yet "enlivened by music and singing."[7] Music supplemented the interior display of landscapes throughout our period, from the eidophusikon of 1781—originally featuring the musical talents of Charles Burney, Jr., Michael Arne and his wife, and Sophia Baddeley—to the *Route of the Overland Mail to India* shown at the Gallery of Illustration on Regent Street in 1850 by Thomas Grieve (ex-scene painter for the Italian Opera) and William Telbin, with music by Michael Lacey. Music-playing automata were also displayed at the Gallery of Practical Science on Adelaide Street (an amusement hall described by Richard Altick as "the first direct English progenitor of the modern science and technology museum"). Merlin's Mechanical Exhibition, a display of "scientific toys" that occupied a house on Prince Street around 1800, featured a fully automated organ "imitating the performance of a full band." After this exhibit closed, a portion of this instrument was incorporated into an immense mechanical orchestra by Flight & Robson called the Apollonicon, installed at the Royal Cyclorama or Music Hall adjoining the Colosseum in Regent's Park from 1848. When the cyclorama reenacted natural spectacles such as the 1755 Lisbon earthquake by means of "an exhibition of movable paintings" or enlarged eidophusikon, the Apollonicon performed the first movement of Beethoven's Pastoral Symphony, as well as excerpts from Mozart's *Don Giovanni*, Auber's *La muette de Portici*, and Rossini's *Mosé in Egitto*. If the goal of such visual entertainments (it has generally been agreed) was "inner experience"—a tracing of pathways in the imagination—their musical records attest rather to an interest in mechanics, automatic movement, and collective animation: an interiority of now-unrecognizable form. The domains of "music" and "science" overlapped here in ways largely lost to us.

ECSTATIC ACCUMULATIONS

We could go on and on in our delirious pursuit of knowledge, piling fact upon fact, making ever-longer lists in order to document everything seen and heard. This would make sense, since, particularly during the second quarter of the century, London's musical and scientific activities were shaped by, and understood in terms of, a metropolitan mania for acquisition. One way to reflect upon the need "to have it all," indeed, would be to admit the extent of our own scholarly appetite for more and more research material. The fantasy of transparent representation, after all, requires "data-rich" analysis and twenty-four-hour access to national libraries, museum collections, internet archives, and Google technology. We would do well to recognize these means by which we accrue academic capital, and to cultivate a disingenuous relation to the dream of absolute description—that is, to parade the rapture of "being in control of all the facts" with tongue firmly in cheek. The point to make is that our own glass houses and history books are haunted as much as those of our nineteenth-century antecedents by fantasies of total possession.

Their imperatives were ours: to hoard knowledge and wallow in the vertiginous experience of accumulation.

One such wallower—and one of the more prominent musical explorers brought to London—was Felix Mendelssohn, who first visited in April 1829. Within a week of arrival, in a letter to his family, the wide-eyed musician described the city as a whirlpool, teeming with unemployment, smog, and noise. Mendelssohn catalogued the "awful mass" thus:

> It is fearful! It is maddening! I am quite giddy and confused. London is the grandest and most complicated monster on the face of the earth. How can I compress into one letter what I have been three days seeing? I hardly remember the chief events, and yet I must not keep a diary, for then I should see less of life, and that must not be. On the contrary, I want to catch hold of whatever offers itself to me. Things roll and whirl round me and carry me along as in a vortex.[8]

His giddy experience of urban compression was intensified by the sight of so many immigrants—Irish, Africans, Indians, Chinese—and the experience of racial and gendered mixing on Regent Street: "Then there are beggars, negroes, and those fat John Bulls with their slender, beautiful two daughters hanging on their arms."[9] The young Berliner imagined himself at the eye of a cosmopolitan centrifuge, a storm drawing to its center rich and poor, young and old, men and women of every nation.

How did this, a musician's view, resonate with those of the "scientist"? For one, the accrual of both scientific and musical expertise depended more and more on collection. According to Bernard Lightman, it was not so much the laboratory as the museum that would become the principal organizing space of the particularly Victorian scientific enterprise.[10] The stockpiling of curious objects would increasingly provide the basic raw material for nineteenth-century musical knowledge, as well as for any other improving pursuit. If not showcased in the genius of the displayer—as in the spectacular scientific and musical demonstrations of Walker, Crotch, Davy, Faraday, and Wheatstone—science itself would increasingly be defined in terms of the ordered displays of London's public collections. Lightman's claims for the importance of the museum echo Lydia Goehr's influential notion that the musical canon (emergent, as noted above, precisely in the period considered in this volume) was an "imaginary museum of musical works," curated in the hallowed space of the concert hall. We observe these developments taking place in less illustrious public spaces of collection, concerned with accumulated knowledge as *supply*: the emporium, the bazaar, the warehouse and storehouse, and the exchange, as well as the institutions of learning and public improvement that were modeled after them.

One such institution was London University (now University of London). In 1833, its senior vice-president boasted in telling terms of his city's worldliness; the occasion was a dinner for seniors and faculty at the Freemason's Tavern on Great Queen's Street, the adjoining hall of which had long since hosted the subscription concerts of the Academy of Ancient Music. He announced that London, the "Empress of the Commercial World," was a city now universally recognized as "the resort of all nations"; and yet, before the founding of University College in 1826, or the newly built North London Hospital (now University College Hospital), London had lacked what he called an "emporium for the supply of intellectual necessities."[11] The nascent university certainly boasted storehouses enough of knowledge. Besides Edward Turner's impressive chemical laboratory, the institution owned a large anatomical teaching museum and Robert Edmond Grant's impressive zoological collections, and had just built a new library to house around eight thousand volumes. The new metropolitan university, which had expanded to include King's College (founded in 1829), was in this sense imagined as a kind of "stock exchange," a place for accumulating capital.[12] The scientific impulse to inquire after wonders, in other words, could be indistinguishable from the imperative to acquire them.

The warehouse could also host accumulations of musical knowledge. Hanover Square and Oxford Circus in particular became infested by "harmonic" depots, as Mayfair's bustling shopping district developed into a center for the burgeoning fortepiano and music publishing trades. Around 1830, repositories for music might include Cramer & Beale's Music Warehouse, Cocks & Co. Music Warehouse on Regent Street, and Chappell & Co. on New Bond Street; a little further away, at Soho Square, came Goulding & D'Almaine's Music Storehouse; further still was Clementi & Co.'s Music Warehouse in Cheapside. Music-circulating libraries traded alongside these emporia, increasing the public dissemination of sheet music and meeting the demand for a new, covered shopping experience for the gentry, one that would evade the rude atmosphere of open-air markets. The most longstanding of these stores were Birchall & Co.'s "Musical Circulating Library" on New Bond Street, Mori & Lavenu's "New Musical Subscription Library," and Boosey & Co's "Foreign Musical Library" on Hollis Street. Like its university, London's foremost venue for opera was praised as a warehouse where the performing riches of Europe were amassed. A contributor to the *Quarterly Musical Magazine and Review* hailed the Royal Italian Opera as a "grand emporium from which the supply of music and its examples are set forth," and only then resorted to astronomical metaphor, lauding it as "the centre from which light is projected, and which the *stellae minores* all move."[13] In sum, then, these institutions were quarters for all knowledge—musical totalities, if you will—and waypoints for merchandise.

Musical "specimens" were hoarded too, for music apparently required incubation to flower in London. French, Chinese, Bohemian, African, Javanese, and Germans could be heard across the metropolis, and every musical personality of European note made temporary residence on the Thames, including Vincenzo Bellini, Hector Berlioz, Angelica Catalani, Frédéric Chopin, Joseph Haydn, Luigi Lablache, Jenny Lind, Franz Liszt, Carl Maria von Weber, Maria Malibran, Gioachino Rossini, Giovanni Battista Rubini, Giuseppe Verdi, and Richard Wagner (in 1855). Italians especially were collected by the boatload. In an 1829 assessment of the King's Theatre, the *New Monthly Magazine* suggested "treating the human exotics from Italy precisely in the same manner as we preserve the botanical specimens imported from that country." These singers were canceling too frequently after a wintery March and several bouts of illness; the critic argued that such snags could be unraveled by "the erection of a spacious human Conservatory, under the roof of the King's Theatre, conveniently fitted up, and tempered by the admission of warm air under the regulation of a thermometer."[14] "The Italian opera is an exotic in this country," the *Court Magazine* agreed a decade later, "a hothouse plant that must be nursed with care, and perfected at no little cost of money and attention."[15] The prominent Belgian music writer François-Joseph Fétis thought similarly in his review of the state of music in London of 1829. Such horticultural magnificence could only be explained by tracing "the reign of foreign musicians in England" to the arrival of Handel, whose model attracted Francesco Geminiani to found a school for violin, and many other residents: Carl Friedrich Abel, Johann Christoph Bach, Muzio Clementi, Johann Baptist Cramer, Jan Ladislav Dussek, Giovanni Battista Viotti, and Domenico Dragonetti. From Handel onward, wrote Fétis, "music in England seemed to resemble certain exotics, which can thrive on a soil different from that which gave them birth only be excess of care."[16] No other European city could rival London's artificial climes and glittering botanical diversity.

In such indoor-outdoor spaces, alongside other captive fauna from exotic locales, a musical canon was incubated. In 1831, Edward Cross opened his menagerie for lions, tigers, and other carnivora under six thousand feet of glass at the Royal Surrey Zoological Gardens; in 1845 the site would also see Louis Jullien's first monster concert, with its three-hundred-piece orches-

tra on a purpose-built platform opposite Burton's giraffe house, and "Suoni
la tromba" from Bellini's *I puritani* scored for a hundred mixed brass, ophi-
cleide, and serpent. Only five years earlier, Jullien had been another pioneer
of that strange class, the indoor "promenade concert"; his concerts at the
Theatre Royal, Drury Lane in 1840 brought indoors the pleasurable open-air
"vocal and instrumental" entertainments so fashionable in the early-century
gardens of Vauxhall, Ranelagh, and Marylebone. And finally, of course, there
appeared the largest of all such "crystallizations" of knowledge: the Crystal
Palace—a "total interior" attempting, in the urban countryside of Hyde Park,
to enclose the world within a glittering hothouse of glass and iron. If music
and its instruments had been a scattered presence at the original Crystal
Palace, they amassed into an aggregate of considerable force when the structure moved to Sydenham in 1854; there, it housed the mass "universal" concert events organized by George Grove, the young civil engineer and future
musical dictionary maker, who had just returned from a cast-iron lighthouse
project in Jamaica.

Perhaps it should not surprise that tigers and giraffes were among the
early attendees to the imaginary museum of musical works. This was, after
all, the era of urban naturalism, pastoral return, and the "greening" of London. The process of bringing the whole countryside to town was led by such
master gardeners as Humphry Repton, who laid out Russell Square in 1800
as the centerpiece of the Duke of Bedford's development plan for northern
Bloomsbury, with its elegant perimeter walk, hedged so as to screen it from
the street.

In later generations, visionaries such as John Nash would be charged
with the "naturalization" of the British capital. In his capacity as architect
in the Office of Woods, Forests, and Land Revenues, an agency of the Crown
answerable to parliament, Nash from 1813 onward oversaw a grand scheme
of "Metropolitan Improvements," a project that incorporated such projects as
the draining of Regent's Park and efforts to manage the western spread of the
city by landscaping Hyde Park, Green Park, and St. James's Park (where the
existing canal was converted into a "natural" lake, and formal avenues were
rerouted to form meandering garden paths). Within the "rural city," Nash's
most notable accomplishment was the conversion of New Street into Regent
Street. Nash's brief was a "royal mile" linking Regent's Park with Carlton
House, the home of the future George IV (demolished in 1825). The thoroughfare was impressively wide, with national and royal landmarks punctuating
its progress. Arcades were erected, arching like repeating biological forms
down to Charing Cross and the steps of the Royal Italian Opera. By 1818, the
year before Regent Street was christened, Nash had overseen an exterior and

interior remodeling of the King's Theatre, only a stone's throw from Carlton House, including a new façade, wrap-around colonnades, and an enclosed Opera Arcade on the west side. (The King's Theatre—Her Majesty's at the accession of Victoria in 1837—was traditionally the venue for Italian opera.) In his quest for the urban picturesque, Nash also cleaned up the area around the Haymarket by the widening of Pall Mall and other measures to separate out the exclusive West End from the slums of Golden Square to the east. By 1828, commentators could think of "few cities, even those of Italy of which we speak as wonders of architectural magnificence, [that] can present a drive equal to that from the Opera House to the Zoological Gardens."[17]

Most interesting for our purposes, such an opening out of the city was achieved not merely by the removal of squalor and the cleansing of over-crowded streets, but by purifying the aural environment. In this sense, Nash's "improvements" sought to re-engineer the political geography of the city by harmonizing the urban surrounds and clearing the air. In *Nicholas Nickleby* (1838–1839), Charles Dickens noted the extent to which undesirable sights and sounds had come to be located in places like Golden Square—"one of the squares that have been," where foreigners and other musicians dwelt. A hidden kind of musical London was only barely betrayed by those "dark-complexioned men who wear large rings, and heavy watch-guards, and bushy whiskers, and who congregate under the Opera colonnade, and about the box-office in the season, between four and five in the afternoon, when they give away the orders." These men, he noted, lived in the "musical boarding houses" of Golden Square, along with many minor members of the opera band. Also inhabiting this Square—only because it was not on anyone's way to or from anywhere—was an underclass of "street bands" and "itinerant glee-singers," the notes of pianos and harps floating about the head of the mournful statue of George II. On a summer's night, one experienced "sounds of gruff voices practising vocal music." There, musical pollution mingled with the olfactory: "snuff and cigars, and German pipes and flutes, and violins and violoncellos, divide the supremacy between them."[18] This "region of song and smoke" represented all that the fashionable meccas of the Burlington and Oxford Arcades displaced—the material conditions that glass and window shopping rendered invisible.

The story we are telling here is, at least in part, one of the progressive "disciplining" of our objects of knowledge, which occurred in tandem with these schemes of urban reorganization. This process had begun an institutionalizing phase during the first decades of the nineteenth century, when the task of offering affordable popular science to the masses was entrusted to a new rank of commercial "Arts and Sciences" establishments, whose speciation has recently been traced by Jon Klancher.[19] These were speculative enterprises that offered public lectures in science, moral philosophy, poetry and drama, business and technology, the arts of printing and engraving, as well the "fine arts" of poetry, music, painting, and architecture. Among these was the London Institution of 1806, where Crotch, Wesley, Taylor, Gauntlett, Bennett, Bishop, and Ella would lecture on music in the shadow of the scientific demonstrations of such luminaries as Dionysius Lardner and Richard Owen. Other establishments, such as the British Institution (1805), Surrey Institution (1808), Russell Institution (1808), and Metropolitan Institution (1823), were vital for opening public access to knowledge. Besides housing lectures and administrative processes, these institutions often built ambitious print libraries, reading rooms, laboratories, and museums. The practices of these institutions serve as a reminder that the caesura between subjective and objective knowledge, between the arts and the sciences, had not yet been prized open, or their entanglement obscured by later disciplinary formations.

As the hegemony of "Royal" scientific establishments overseen by Sir Joseph Banks (1743–1820) weakened, and the Royal Society entered a period of decline, Londoners stood before an ever-widening vista of domains. The so-called romantic sciences might include geology, anthropology, chemistry, biology, physics, and other elite disciplines that we now recognize as our own; but also ones that we would not, such as eschatology, phrenology, and mesmerism. The metropolitan embrace of quantity was perhaps nowhere more striking than in the remarkable expansion of the statistical sciences in the 1830s: political economy, the social sciences, criminology, public health. The mania for "facts, nothing but facts" was fed by the founding of Babbage's Statistical Society of London in 1834. During the 1830s, popular science came to be a vocation for "societies" of reformist bent. The call was answered in 1836 by the institution of the Society for the Diffusion of Useful Knowledge, and again two years later by the Polytechnic on Regent Street, a venue where patrons could attend a harp recital, a lecture on chemistry, and then dissolving views, all attesting to "the advancement of practical science."

Musicology too was a feature of this new disciplinary landscape, as, toward midcentury, there emerged a host of institutions devoted exclusively to cultivating music as both a form of reproducible "science" and archivable "literature." The appearance of long-running specialist music magazines and a thriving culture of music criticism, particularly in the 1820s and 1830s, owed much to the activities of these institutions and publishing houses. News and arts weeklies with strong musical concentrations abounded. Here are more lists: the *Examiner* (1808), *Atlas* (1826), *Athenaeum* (1828), and *Spectator* (1828) appeared alongside general magazines such as *Blackwood's Magazine* (1817), *London Magazine* (1820), the *New Monthly Magazine* (1821), and review journals like the *Foreign Quarterly Review* (1827) and the *Westminster Review* (1824). Early issues of the specifically music-themed *Harmonicon* (1823–1833), for example, devoted unusual energy to reviewing sheet music published by the Harmonic Institution in ways that suggest a less-than-innocent relationship with both the Argyll Rooms and the Philharmonic Society. The *Harmonicon* was a hybrid publication, mixing "original papers on every subject that can interest the Musical World"—in the style of the *Quarterly Musical Magazine and Review* (1818–1828)—with the usual catalogue of engraved sheet music in the mode of eighteenth-century serial collections. Its first number declared its intention to make "stores of music" available to "thousands of amateurs." "The intelligent admirer of the science," the writer explained, demands the "rare union of Literature with music, of which the *Harmonicon* will become the medium." The *Harmonicon's* editor-in-chief, William Ayrton, was a founding affiliate of the Royal Institution, a member of the Philharmonic Society, and ex-director of the Italian Opera, as well as a chief consultant in Nash's Argyll Rooms project.[20] The year 1851 saw the emergence of the Musical Institute of London, established for "the cultivation of the art and science of music"; to an unprecedented degree, this seems to have entailed reading, writing, and talking about music, as the founders of the institute planned for "a library of music and musical literature, and a museum, and the provision of a reading-room; for the holding of *conversazioni*, for the reading of papers upon musical subjects, and the performance of music in illustration; and for the publication of transactions."[21]

Training in musical performance was also increasingly the business of institutions. Instructional schools—conservatories proper—proliferated, and many came and went, such as Jullien's Royal Conservatory of Music (1845–1858), which was attached to his publishing house and circulating library on Regent Street. Of more lasting fame was the first public teaching institution devoted to "the science and practice of music": Lord Burghersh and Nicolas-Charles Bochsa's Royal Academy of Music. The academy's founding mission of 1822 was to "to promote the cultivation of the science of music, and, by affording the means of obtaining perfection in this branch of the fine arts, to enable the natives of this country who are desirous of obtaining a knowledge of this science, to provide for themselves the means of an honourable and comfortable livelihood."[22] The charter, in this sense, closely aligned with Nash's project to renovate the metropolis. Modeled on the Royal Academy of Painting, the Royal Academy of Music opened a few steps from the Hanover

Square Concert Rooms, the venue famously conceived as London's mecca for music, balls, and masquerades, opened some forty years earlier by Johann Christian Bach and Carl Friedrich Abel.

The changing fortunes of the Assembly Rooms at Hanover Square may reveal much about shifting institutional practices and configurations of musical knowledge. Activity on Hanover Square had calmed since the glory days of 1774, though as this volume attests, the area maintained strong musical and scientific associations well into the nineteenth century. The arrival of the Society of the Concerts of Ancient Music in 1804 brought with it a strict injunction against any compositions less than twenty-five years old. Led by musician, astronomer, and mathematician Thomas Greatorex, these "Antient" events took place sans candles or lusters. Instead, the interior was illuminated by "transparent paintings" by Gainsborough, West, and Cipriani, which were lit from behind. When the Royal Academy was founded, these paintings were taken down. If one commentator suggested that womanly vanity was to blame—"The variegated hues were so unfavourable to the complexions of the female part of the audience, that the paintings were necessarily, though reluctantly removed"—it seems more likely that such displays, which divided spectatorial attention between framed transparencies and recital numbers (in the manner of the eidophusikon), were beginning to seem dated.[23] The Assembly Rooms continued to host a great variety of spectacles, but increasingly specialized ones. From 1833, the venue welcomed the concerts of the Royal Philharmonic Society: Mendelssohn conducted the English premiere of his own Scottish Symphony there on June 13, 1842. Scientific talks also graced these rooms: James Braid's lecture on neuro-hypnotism (March 1, 1842) was followed two days later by a "Quartet Concert" featuring Mozart's Piano Quintet in C Major (KV 415), Mendelssohn's "Frülingslied," and a string quartet by Beethoven.[24]

During the 1820s, the Assembly Rooms came into direct competition with the Argyll Rooms, a new model venue for enclosing music *en masse*, situated at the intersection of Regent and Little Argyll streets just a stone's throw to the east of Hanover Square. As Leanne Langley has described, this high-profile performance space was built in conjunction with a joint-stock company calling itself the Regent's Harmonic Institution, a select group of "professors" with strong links to the now well-established Royal Philharmonic Society, which was formed to enhance the status of instrumental music in the capital.[25] Its music room, lined with fluted Corinthian pilasters, was the largest of its kind in London, seating eight hundred spectators (the Assembly Rooms at Hanover Square were full at five hundred). It was conceived as a kind of artificial conservatory, ventilated by the novel use of air-conditioning, while a "reservoir for gas" or "gasometer" made it among the first buildings in London to boast interior gas lighting (another was the Theatre Royal at Covent Garden). On the ground level, the building presented a glass-paned commercial frontage of four immense windows and a covered walkway enclosed by decorative balcony and an iron railing to shield it from the street. This dazzling construction separated patrons from Nash's thoroughfare and gave intimate access to a "Music Shop or Warehouse." The latter, a "museum for selling music along the front of the street," provided a useful foil to the grand, parallelogram-shaped "music room and theatre for French plays" on the upper floor. The Argyll Rooms would host an illustrious array of events, including the first rehearsals with a baton conductor (Louis Spohr in 1820), the debut of the twelve-year-old Franz Liszt in 1824, the first English performance of Beethoven's Ninth Symphony in 1825 (a work commissioned by the Royal Philharmonic Society), and Mendelssohn's celebrated 1829 displays.

The Argyll Rooms were maintained in tandem with an ambitious music-publishing initiative. An 1819 advertisement announced the founding of the Harmonic Institution "upon the most extensive and liberal scale, for the purpose of printing, publishing, and vending Music, [and] for the sale of Instruments." In addition to buying and selling, the announcement declared plans for a commodious subscription library stocked with "all classical compositions in every branch of science." These would be "methodically arranged" alongside "Histories, Disquisitions, Treatises, etc." The institution even promoted something like an early Urtext ideal, promising "elegant editions of the best musical works, and to guarantee that each shall not only be free from the too usual errors of the engraver, but that it shall be published in conformity to the Author's intention."[26] One such author was again Beethoven, whose formidable "Hammerklavier" Sonata, op. 106, was published within the first year of the institution's establishment in the composer-authorized two-part version. Such "classical" music, printed for posterity "upon superior and durable paper," was necessarily of the most permanent and serious kind. As has been noted, the commercial activities of this and similar institutions point ahead to the crystallization of a musical canon, although that canon— as any browse through the Harmonic Institution's catalogue makes clear— hardly resembles later, more modern formations. One irony of the institution's quest for permanence is that the Argyll Rooms soon burnt down, as did so many of London's repositories for music. Rumor had it that the conflagration was caused by one Ivan Chabert, "The Fire King," who for two years had used the venue for exhibiting his prowess in such acts as swallowing phosphorus and entering a heated oven for the purposes of cooking a steak.

No sooner had the Argyll Rooms combusted than another, even larger multipurpose concert venue emerged: Exeter Hall, built on the Strand between 1829 and 1831 to host crowds of thousands. In Exeter Hall, the aggregate forces of music and science alike were put in service of social improvement—and this time the wild animals were cleared away first. It was built on the former site of Edward Cross's Royal Grand National Menagerie, where flâneurs and natural historians alike studied a Bengal tiger, hyena, lion, jaguar, sloth, camel, monkeys, hippopotamus, rhinoceros, elephant, ostrich, cassowary, pelican, "emews," cranes, eagle, cockatoos, elks, kangaroos, antelopes, and other exotic species.[27] Exeter Hall played host to a similarly dazzling variety of reformist species, united only by an interest in social improvement and a need for an indoor meeting place as large as a cathedral but without its godly specificity. For Methodists, evangelicals, enthusiasts, philanthropists, and other nonconformists, this meeting place was the center of the religious world. Its committee rooms, offices, and Great Hall hosted meetings great and small of the Protestant Association, Temperance Society, the Anti-Slavery Society, Ragged School Union, Bible Society, Wesleyan Missionary Society, London Missionary Society, Young Men's Christian Association, and more. Exeter Hall also hosted scientific lectures aimed at improving the conditions of the masses, as did, for example, David Boswell Reid's 1842 lectures, "The Chemistry of Daily Life," wherein he educated the population in questions of gaslight, heat, sanitation, hygiene, and overpopulation, and Edward Davy's 1839 exhibition of an electric telegraphic apparatus to rival the Wheatstone prototype discussed in the present volume.

And it provided a place in which to cultivate the hygiene of Great Music, for those who wished to profit from its salutary power. One such profiteer was the music publisher J. Alfred Novello, who would later pioneer ways of introducing inexpensive sheet music to "the million." His house organ was *The Musical World, a Weekly Record of Musical Science, Literature and Intelligence* (1836–1891), a vociferous advocate for the moral and intellectual value of masterworks. A brief manifesto appeared in the second year of its existence, where the author (probably Henry Gauntlett, who often lectured for the Islington Literary and Scientific Institution) noted that accessibility to the "monuments of departed genius" had finally equipped men to distinguish good music from bad. "The immense accumulation of clasical [sic] music, the frequency of its performance, and its wide dissemination," the writer claimed, "have been rendering and will continue to render us, more and more fastidious."[28] Working out of his "music warehouse" in Soho, Novello was keen to position this serious weekly at the center of a global marketplace for music. His hope was to "advance" musical taste, enlist music as a vehicle for the improvement of all classes of men, and amass money.[29] Nowhere was this better expressed than in Novello's decision to purchase Joseph Mainzer's *Musical Times and Singing Circular* in 1844. Since the banning of similar efforts with the laboring classes in Paris, Mainzer had taught adult classes in singing at such dissenting venues as Mechanics' Institutes across London and Westminster, the national schools of St. George's, Hanover Square, and other parishes.

From 1834, Exeter Hall was the site of the Sacred Harmonic Society's Amateur Musical Festivals, in which Novello was an enthusiastic participant. These festivals, which pooled the forces of all the minor choral associations of the metropolis, made this "Exeter Hall Society," in the words of Novello's *Musical World*, "a point of centralization." Comparisons to natural phenomena were characteristically abundant: Exeter Hall was "an all-absorbing focus, which attracted every minor light," and "a majestic ocean, receiving every tributary stream." In 1839, the venue acquired the first purpose-built modern concert organ. "Imagine a gigantic hall with places for 3,000 persons, crammed full, head above head," Spohr wrote in his autobiography, with a "magnificent and stupendous organ, and on all sides around it, an orchestra and choir of singers numbering 500 persons."[30] John Hullah, meanwhile, ran his famous "monster classes," "musical evenings," or "choral meetings" for "the instruction of schoolmasters in singing" and Sunday schools. It was in the Great Hall in 1842 that Mendelssohn played J. S. Bach's Fugue in E-flat, where the same composer oversaw performances of his oratorios *St. Paul* (1837 and 1844) and *Elijah* with Jenny Lind (1847), and where (in 1852) Berlioz conducted the orchestra of the New Philharmonic Society in Beethoven's Ninth Symphony. For the bulk of its visitors, then, whether they were participants in scientific or musical innovation or in some other improving enterprise, Exeter Hall was conceived as the moral center of empire, bringing together the voices of creation and gathering news from "the sons and daughters of earth thousands and thousands of miles away."[31] It survived in this form until 1907, when the Strand Palace Hotel was erected in its place; the latter stands there now, its famed art-deco façade cut away in favor of sliding doors of reflective glass, its red-velvet upholstery still welcoming weary travelers in the comfortable embrace of faded imperial glory.

Less successful than Exeter Hall—and with the imperial nature of its project even more tellingly on display—was the Royal Panopticon of Science and Art, a vast venue in the model of a mosque, which opened on Leicester Square in 1854. The Royal Panopticon was built for the purposes of demonstrating the latest discoveries of science and manufacturing "in a popular form." Music was performed here too, but as one arm of the science of acoustics—its founding documents attest—and these educational entertainments featured alongside "pictorial views and representations" illustrating "history, science, literature and the fine and useful arts." The Panopticon was suitably equipped with an enormous pipe organ commissioned from William Hill, featuring three separate consoles and numerous "mechanical novelties," including drums, crescendo pedal, and vibrato. The organ was the largest in the world, its makers claimed. Indeed, such was the Panopticon's monumentalizing thrust that its various musical and visual entertainments could barely coexist even under its cavernous dome: the organ itself had a large hole in its middle to accommodate the projector for the "dissolving views."[32] The exterior was decorated with iconography representing a suitably diverse array of national talents, including Henry Purcell, Isaac Newton, James Watt, Humphry Davy, William Shakespeare, Francis Bacon, and William Herschel. Its shareholders soon found, however, that one could no longer bank on a market for such promiscuous pan-disciplinary display; they had to sell, and quickly, after a mere two years. One commentator used the demise of the Panopticon as an occasion to meditate on the decline of popular science more broadly:

> There were, doubtless, many who, at the time, regretted the conversion of a magnificent and unique Hall of Science into a palæstra of unblended amusement; but where was their practical loyalty to the cry of "to the rescue" in the time of its death throes? . . . Numerous attempts have also been made to cultivate a taste for abstract science by attractive experiments and engaging illustrations, [but] these have culminated in the discovery that the nearest approach to success has been when the greatest amount of recreative amusement was presented.

Among London's once-mighty institutions of popular science, the commentator noted that only the Polytechnic had "remain[ed] partially true to its traditions," while still falling victim to the "prevailing disposition for sensational effect"; and the Mechanics' Institutes had become the forum merely for "abortive attempts at theatrical elocution, and poetical and facetious readings."[33] In the midst of this depressing social and moral decline, the commentator declared himself positively refreshed to see the Royal Panopticon of Science and Art ultimately refurbished as the Alhambra Palace and Music Hall, home to tumblers, aerial acts, ballets, equestrian shows, dancers of the can-can, and hedonists galore.

Centered for a moment on Regent Street, on Exeter Hall, on Hyde Park, and then everywhere else, London's musical and scientific life was unrivaled for sheer volume and range of attractions. Music spilled out onto the street, as Berlioz noticed near the end of our period. "There is no city in the world, I am convinced, where so much music is consumed as in London," he wrote in 1851, having visited a Chinese junk moored on the Thames, where he listened to two destitute Indians from Calcutta, and heard three blackface "Abyssinians" on violin, guitar, and tambourine. "It follows you even in the streets, and that is sometimes not the worst of its kind."[34] The ecstatic accumulation of music extended to foreign musicians, subscription libraries, conservatories, instrumentaria, parlors, and botanical gardens; it encompassed scores, periodicals, monographs, and—most of all—events: concerts, operatic entertainments, musical meetings, choral society gatherings, street music scenes, scientific performances, and crowds. Crowds were everywhere, attending a bewildering array of exhibitions, church services, lectures, demonstrations and nonevents: from Michael Boia at the Egyptian Hall in 1831 (who made music by hitting his chin—one *pièce de résistance* was Cherubini's overture to *Lodoïska*), to Joseph Richardson and Sons' "Original Rock Band" at Stanley's Rooms in 1842, to the first performance on May 4, 1795, of Haydn's final "London" Symphony in D Major, no. 104, directed by the composer himself in the then brand-new concert room adjoining the King's Theatre. The hunger for edification and entertainment was insatiable, as London's diverse throngs consumed music with a passion and unpredictability that, in the end, gives lie to any conception of this accumulation of people as some undifferentiated "mass."

AN ACCOUNT OF THIS COLLECTION

Our volume is organized roughly by chronology and framed by questions of epistemology. We begin by wondering what kind of object music was understood to be in the last decades of the eighteenth century, and end by asking this again about music circa 1850—and in both cases we find that something may be learned by studying the ways that music partook of scientific methods and technologies. Between these brackets is necessarily something of a cabinet of curiosities: no musical repertoire, practice, or field of inquiry is assessed in its entirety; for that, the reader is referred to the wonderful specialist studies by Simon McVeigh, William Weber, Deborah Rohr, David Wyn Jones, Susan Wollenberg, Ian Taylor, Ian Woodfield, and others. Rather, the authors gathered here consider a series of objects for which the lenses of disciplinarity have necessarily been inadequate: which have, indeed, largely eluded their notice.

Emily Dolan argues that Burney's *General History of Music* (1776–1789) marked a major epistemological shift: a change in the kind of "object of knowledge" music was understood to be, bringing along with it new methodologies for its study and classification. Burney sought to create a comprehensive account of music not by means of mathematics-based theory (as in Padre Martini's *Storia della musica* of 1757–1781), or biography (as in Vasari's model), but rather through first-hand observation of musical performance and its instruments. Dolan suggests an affinity between Burney's *History of Music* and contemporaneous endeavors in the natural sciences, including his own writings on astronomy, and the taxonomies of his friends Joseph Banks and William Herschel. While on his tours (as Dolan notes), Charles Burney was an enthusiastic witness to the demonstration not just of musical technologies, but scientific ones as well; and in his travel diaries, instruments like C. P. E. Bach's Silbermann clavichord bump up against Philip Matthäus Hahn's orrery in Ludwigsburg. and Roger Boscovich's *Stet Sol*, a prismatic device for studying sunlight.

In the next chapter in our volume, Deirdre Loughridge shows how musical and scientific instruments might feature alongside one another in London's culture of amusements during the years around 1800. Loughridge considers the career of Adam Walker, a self-taught practitioner of "natural and experimental philosophy" from the north of England who gave public lectures and demonstrations on astronomy and the natural sciences. "Celestial Mechanisms" focuses on two of Walker's inventions that often featured together: the celestina, a harpsichord-like keyboard instrument whose strings were sounded with a bow; and the eidouranion, a vast machine that provided views of astronomical phenomena, scaled to London's largest theaters. Loughridge notes that while the celestina's continuous, unearthly tone and rotating inner mechanisms derived from an older notion of celestial harmony, the instrument was also marketed domestically as a harpsichord stop, particularly suitable for the performance of slow movements.

The eidouranion—meaning image (or imitation) of the heavens—was one of a handful of eido-entertainments to appear in London during the 1780s; as Loughridge notes, Walker surely modeled his new "transparent orrery" on Philippe-Jacques de Loutherbourg's eidophusikon, a similarly mobile animated instrument that presented scenes of natural phenomena such as sunrises, shipwrecks, and exploding volcanoes. Both the eidophusikon and the eidouranion created their elaborate illusions by means of concealed mechanisms, elaborate lighting tricks, and images projected onto screens from painted transparencies; both entertainments featured music. Walker and de Loutherbourg alike organized their demonstrations of their instruments in numbered series of views, which were described meticulously on the play-bills.

This serial arrangement served as a model to the London composer Nicola Sampieri in his "Concert upon an Entire New Plan," given at the Hanover Square Rooms in 1798, and elsewhere in the British Isles during the following decade. Sampieri's career is considered in Ellen Lockhart's essay. As in the eido-entertainments, Sampieri's "New Plan" employed "numerous and beautiful transparencies" to present a series of natural scenes, both mundane and meteorological. But where music served an ancillary function for Walker's and de Loutherbourg's shows, for the "Concert on a New Plan" it had been composed expressly to represent these natural phenomena. Sampieri's compositions for the fortepiano encouraged listeners to experience a perfect correspondence between what they saw and what they heard. Lockhart suggests that Sampieri's endeavors proposed a new kind of analogy between sights and sounds represented. She notes, further, that Sampieri's music—parts of which were published with dedications to the Lovers of Science—emerged simultaneously with a renewed scientific interest in the analogy between tones and colors, thanks to the experiments of Thomas Young, which demonstrated that light and sound behaved according to a single principle of wave-based movement. Sampieri's work connects the pictorial entertainments of eido-crazed London with the natural and cosmological worlds of such familiar staples of the nineteenth-century symphonic repertoire as Mendelssohn's *The Hebrides*, or even Gustav Holst's *The Planets*. At the same time, Lockhart's project is to rebalance the long-held notion of this culture as visual-centric, attesting instead to a broad interest in engaging the eyes and the ears together, and an investment in analogical thinking both within art and within science.

For Thomas Young, an early expertise in music and its instruments led directly to a study of acoustics and the properties of sound, which gave rise in turn to mature contributions to an extraordinarily diverse range of inquiries. Such a career trajectory was also enjoyed by Charles Wheatstone. In chapter 4, Myles W. Jackson argues that Wheatstone's achievements in the fields of acoustics and long-distance communication were the direct result of his early education in musical instrument building. Jackson begins by considering the scientist's early career in light of the particular culture of musical and scientific showmanship based on and around the Strand, where his father had a musical instrument shop. Jackson takes us from exhibitions of "Charley Wheatstone's Clever Tricks" through the kaleidophone and other devices investigating sound-light analogies and the wave properties of sound. From there we move to Wheatstone's early experiments in resonance, in which the violin bows, flutes, bassoon reeds, and tuning forks that inhabited the Wheatstones' shop became quite literally instruments of science, generating new acoustic knowledge. Jackson concludes by considering Wheatstone's work during the 1830s on reeds and reed pipes, which resulted not only in an early speaking machine (observed with great interest by a young Alexander Graham Bell), but also in his most famous musical invention, the concertina.

Melissa Dickson makes Wheatstone's enchanted lyre, or "acoucryptophone," the point of departure for a study of fantasies about the materialization and sight of sound. The acoucryptophone was one of the devices on display in the young Wheatstone's "Musical Museum." It consisted of a lyre suspended from the ceiling by a brass wire, which connected it to other musical instruments in the room above; when these were played, their sounds would seem to emanate directly from the lyre. In view of acoustic science, the device demonstrated the principle that sound traveled more effectively through metal than through air. But the case of the enchanted lyre allows us to see how, in early nineteenth-century London, objective demonstration was elided with discourses of the supernatural. The scientific proof was also a conjurer's act. Dickson considers how popular-scientific devices like Wheatstone's enchanted lyre and invisible girl—which simultaneously rendered sound waves newly tangible and displaced the labor of performance—may complicate Richard Leppert's notion of music as abstract sound produced by socially constructed bodies.[35] Ultimately, Dickson argues that the acoucryptophone was akin to Wheatstone's telegraph, which also transmitted sound waves across a distance. Thus, in the broader history of technologies of communication, its speciation can be understood to supply a missing link between Orpheus's lyre and the telephone.

In the next chapter James Q. Davies takes this analogy as a point of departure. Opening with a veritable menagerie of musical instruments as communication devices—a species he calls "contact instruments"—he considers the ways in which such devices invented in Victorian London, by Wheatstone and others, were put to use in British colonial encounters. He traces the search, among "non-conformist scientists, music theorists, reform-minded London evangelicals and missionaries," for an Instrument of Instruments, a device sensitive to all non-Western scale systems, and a notational system capable of recording all known vocal sounds. Here we meet Charles Wheatstone in academic costume, as professor of experimental philosophy at King's College London; Davies relates Wheatstone's invention of the concertina in 1829 to his interest in free-reed speaking machines and the resonance of the Javanese *gendèr*.

Chapters 7 and 8 are a pair devoted to music and energy: Sarah Hibberd suggests that musical sound could be seen to represent human *force vitale* on the stage, while Gavin Williams shows how sound could be perceived as a threat to the productivity of the urban workforce. Hibberd considers two stage adaptations of Mary Shelley's *Frankenstein* that appeared in London in 1823: Richard Brinsley Peake's *Presumption; or, The Fate of Frankenstein*, which played at the English Opera House near the Strand, and Henry Milner's *Frankenstein; or the Man and the Monster!*, for the Coburg, a theater for a primarily working-class audience, which was situated on the South Bank of the Thames. Both of these plays were melodramas, which meant that they featured both spoken dialogue and mime, the latter conventionally accompanied by descriptive music. In each of Peake's and Milner's productions, the monster was mute. Hibberd situates Shelley's novel and its adaptations in the context not only of contemporaneous developments in electrical medicine, but also of public debates on vitalism. These debates, which played out in the second half of the 1810s, were oriented in relation to a dispute between two prominent members of London's Royal College of Surgeons.

Gavin Williams shows us Charles Babbage as he neared the end of his life: living on Dorset Street in Marylebone alongside an old "difference engine" and an incomplete "analytical machine," one working automated dancer (rescued from the remnants of Merlin's Mechanical Exhibition), and—by Babbage's own furious account—many hundreds of noisy and disruptive street performers. Williams posits a connection between what he calls "Babbage's favored geriatric occupations": continued work on the "difference engine," and continued campaigning for increased legal restrictions on street musicians. This is a history of listening that draws at once upon Babbage's early designs for the "difference engine" (which required its operator to count the pealing of multiple bells) and the scientist's pamphlet "On Street Nuisances" (wherein he claimed that itinerant musicians had destroyed "one-fourth part of [his] working power."). Williams describes how sound was embroiled in calculations of labor value, and how auditory concerns shaped Victorian notions of artificial intelligence.

Flora Willson's essay concludes our volume by returning to the question of music's epistemological status: what kind of object of knowledge it was believed to be, what material artefacts could speak for it, and in what kind of museum. The year is now 1851—exactly eighty years, almost to the day, since Burney published his first volume on *The Present State of Music*. The occasion is the original Great Exhibition of the Works of Industry of All Nations at the Crystal Palace in Hyde Park. Not unlike Burney, the organizers of the Great Exhibition sought to represent "the Present,"—ordered, classified, and ranked—as the culmination of a historical narrative of progress. In Willson's reading of the Great Exhibition, as in Dolan's of Burney's *General History*, music was represented by its instruments, and partook of their status as objects; the Exhibition featured pianos, organs, violins, and more, as well as internal mechanisms and other component parts, as representative Works of Industry. However, as Willson argues here, although its material traces were scattered throughout the building (classified as "Machinery," "General Hardware," instruments of "Fine Art" and even "Manufactures from Animal and Vegetable Substances"), music was an ever-elusive presence at the Great Exhibition. Its tendency was to recede into the buzz of the crowd, to dissipate, half-unheard, into the towering domes of the steel-and-glass cathedral. Accounts attest that these instruments were played (or "demonstrated," to use the official language), but nowhere is it recorded what was played, or by whom. Indeed, Willson suggests—citing Marx's dictum that in a commodity culture "all that is solid melts into air"—that the Crystal Palace seemed able to render its musical objects immaterial. Even the near-eight-hundred-strong performing force, an army of instruments and voices gathered for the opening ceremony, was only faintly perceived.

And so we end with Marx, having begun with him too. As one of the most charismatic figures in the metropolitan crowd assembled here, his presence in these pages is perhaps overdetermined, his insights so alluring as scarcely to show their age even after more than a century and a half of continual exposure (as attested by the above-quoted passages from the *Economic and Philosophical Manuscripts*). Yet we are not so spellbound as to allow him the last word in the present opening remarks, notwithstanding the fact that they have distended into Total Description (Burney would have sympathized, as would the organizers of the Industrial Exhibition). After all, if the diverse arguments of our volume may be gathered into a single thesis, it is that the visual, the aural, and the tangible mattered together in this London, and continued to matter together throughout our period, despite shifting disciplinary configurations and other dark bourgeois forces. For our purposes, a more apt valediction may be had from Martin Meisel, from the opening of his landmark study of visual and representational culture in London during this same period. The impetus to trace connections between multiple fields of activity, and assess their common elements—Meisel reassured his future interlocutors—is more than merely the curse to be borne by overly acquisitive or materialistic historians. Rather, the indiscipline of the scholar's gaze aimed to follow the situation of the historical metropolis and its agents, reflecting the imaginative and professional scope of period activity.[36] We hope, finally, that the studies gathered in these pages inspire musicologists and sound studies scholars to pay more attention to *Das Land ohne Musik*, and that these studies convince historians of science that music, sound, and hearing were germane to the production of knowledge in this phase of most extraordinary acceleration.

NOTES

1. Karl Marx, *Economic and Philosophic Manuscripts of 1844*, trans. Martin Mulligan (New York: International Publishers, 1968), 140.

2. The great mineralogist, moral philosopher, and polymath William Whewell famously coined the word "scientist" in 1834. The earliest printing of the English word "musicology" appeared in relation to an esoteric dispute over the "dissonant fourth" in the theory of an obscure Blackheath organist called Edward Clare. The German equivalent was used as early as 1822 by the Dublin-based educationalist and alleged quack, J. B. Logier. What we mean by the "discipline" is clearly very different from a nineteenth-century usage; see Johann Bernhard Logier, *System der Musikwissenschaft und des musikalischen Unterrichts* (Berlin: Wilhelm Logier, 1822).

3. The *Oxford English Dictionary* traces the first use of the phrase "the aural" (as opposed to the earlier term "the auricular" meaning "pertained to the ear") to famed philosopher George Henry Lewes's discussion of "sensuous knowledge" in 1847.

4. This extensive literature includes Joe Kember, John Plunkett, and Jill Sullivan, *Popular Exhibitions, Science and Showmanship, 1840–1910* (London: Pickering and Chatto, 2013); Bernard Lightman, *Victorian Popularizers of Science: Designing Nature for New Audiences* (Chicago: University of Chicago Press, 2007); Bernard Lightman and Aileen Fife, *Science in the Marketplace: Nineteenth-Century Sites and Experiences* (Chicago: University of Chicago Press, 2007); David N. Livingstone and Charles W.J. Withers, *Geographies of Nineteenth-Century Science* (Chicago: University of Chicago Press, 2011); Luisa Calè and Patrizia Di Bello, eds., *Illustrations, Optics and Objects in Nineteenth-Century Literary and Visual Cultures* (Basingstoke, UK: Palgrave Macmillan, 2010); Brenda Weedon, *The Education of the Eye: The History of the Royal Polytechnic Institution 1838–1881* (London: University of Westminster, 2008); Jonathan Crary, *Techniques of the Observer: On Vision and Modernity in the Nineteenth Century* (Cambridge, MA: MIT Press, 1990).

5. Benjamin Steege, *Helmholtz and the Modern Listener* (Cambridge: Cambridge University Press, 2012); Veit Erlmann, *Reason and Resonance: A History of Modern Aurality* (New York: Zone Books, 2010); Georgina Born, ed., *Music, Sound, and Space: Transformations of Public and Private Experience* (Cambridge: Cambridge University Press, 2013); Jonathan Sterne, The *Audible Past: Cultural Origins of Sound Reproduction* (Durham, NC: Duke University Press, 2003).

6. The quotation is from Bernard Lightman, "Refashioning the Spaces of London Science: Elite Epistemes in the Nineteenth Century," *Geographies of Nineteenth-Century Science*, 28.

7. John Henry Brady, *A New Pocket Guide to London and its Environs* (London: John W. Parker, 1838), 118.

8. Sebastian Hensel, *The Mendelssohn Family (1729–1847) from Letters and Journals*, trans. Carl Klingemann, 2 vols. (London: Sampson Low, Marston, Searle and Rivington, 1882), 1:178.

9. Ibid.

10. Lightman, "Refashioning the Spaces of London Science," 25–50.

11. "University of London," *London Science and Medical Journal* 1, no. 7 (March 17, 1832): 210.

12. The words "stock exchange" were first incised some thirty years earlier above the entrance of William Hammond's new building for commercial trading on the site of Daniel Mendoza's boxing saloon, near the Royal Exchange.

13. "Sketch of the State of Music in London, May 1822," *Quarterly Musical Magazine and Review* 4, no. 14 (1822): 241.

14. "Music," *New Monthly Magazine and Literary Journal* 27, no. 100 (April 1, 1829): 153.

15. "The Drama," *Court Magazine and Monthly Critic* 10 (1837): 198.

16. François-Joseph Fétis, "M. Fetis on the State of Music in London," *Harmonicon* 1 (1829): 183.

17. "Private Session of the Bills 1828," *London Magazine*, September 1828, 151.

18. Charles Dickens, *The Life and Adventures of Nicholas Nickleby* (London: Chapman and Hall, 1839), 5–6.

19. Jon Klancher, *Transfiguring the Arts and Sciences: Knowledge and Cultural Institutions in the Romantic Age* (Cambridge: Cambridge University Press, 2013).

20. Ian Taylor, *Music in London and the Myth of Decline: From Haydn to the Philharmonic* (Cambridge: Cambridge University Press, 2010), 144.

21. J. W. Davison, *From Mendelssohn to Wagner: Being the Memoirs of J. W. Davison*, ed. Henry Davison (London: W. M. Reeves, 1912), 126.

22. William Wahab Cazalet, *The History of the Royal Academy of Music* (London: T. Bosworth, 1854), 246.

23. "Hanover-Square Concert Rooms," *Manchester Iris: A Literary and Scientific Miscellany* 1, no. 31 (August 31, 1822), 243.

24. "Theatres," *Times* (London), March 5, 1842.

25. Leanne Langley, "A Place for Music: John Nash, Regent Street and the Philharmonic Society of London," *Electronic British Library Journal* (2013), http://www.bl.uk/eblj/2013articles/pdf/ebljarticle122013.pdf.

26. "To the Musical World," *Morning Chronicle* (London), January 15, 1819.

27. This site is now the location of the Strand Palace Hotel.

28. *Musical World* 8, no. 119 (June 21, 1838): 125.

29. Victoria L. Cooper, *The House of Novello: Practice and Policy of a Victorian Music Publisher, 1829–1866* (Farnham, UK: Ashgate, 2003).

30. Louis Spohr, *Louis Spohr's Autobiography, translated from the German* (London: Longman, 1865), 256.

31. "Exeter Hall Pets," *Punch, or the London Charivari*, June 1844, 210.

32. Nicholas Thistlethwaite, *The Making of the Victorian Organ* (Cambridge: Cambridge University Press, 1990), 205–6.

33. William White, *The Illustrated Handbook of the Royal Alhambra Palace, Leicester Square* (London: Nicholls, [1869]), 11–12.

34. Hector Berlioz, "Au Rédacteur," *Journal des débats* (July 29, 1851): 2.

35. Richard Leppert, *The Sight of Sound: Music, Representation, and the History of the Body* (Berkeley: University of California Press, 1993).

36. Martin Meisel, *Realizations: Narrative, Pictorial, and Theatrical Arts in Nineteenth-Century England* (Princeton, NJ: Princeton University Press, 1983), 5.

CHAPTER ONE

Music as an Object of Natural History

EMILY I. DOLAN

They sung their strains in notes so sweet and clear
The sound still vibrates on my ravished ear.
—Dante, translated by Charles Burney

CHARLES BURNEY'S TRAVELS

We begin with an echo. On July 21, 1770, Charles Burney laughed and shouted; he instructed a trumpet to be blasted and a pistol and musket to be fired. His rowdy noisemaking was part of a sonic investigation: Burney was stationed in the outskirts of Milan, at the Villa Simonetta, built in the mid-sixteenth century for Ferdinando Gonzaga. A visitor who went to a particular window on the top floor, one that looked out over the courtyard, could experience a remarkable effect: single sounds reverberated loudly thirty or even fifty times. Athanasius Kircher first discussed the echo in the ninth book of his *Musurgia universalis* (1650) and later returned to the topic in his *Phonurgia Nova* (1673), explaining in detail how the architecture created the conditions for this phenomenon. The echo gained fame over the course of the eighteenth century, and by the nineteenth century the villa was a must-see destination. Burney's visit was just one stop of many during his travels in France and Italy, which he had begun in June 1770, to gather material for a forthcoming *General History of Music*. He published an account of his travels in 1771 and the following year undertook further travels through Germany and the Netherlands, publishing a second report in 1773. The first of the four volumes of his history appeared in 1776; the final volumes were published in 1789.[1]

Burney admitted he had "expected exaggeration" when he visited the echo but had been wrong. He did not attempt—as Kircher had—to explain the physical basis for the phenomenon ("This is not the place to enter deeply into the doctrine of reverberation," he wrote). Rather, he was concerned with sonic effect, declaring, "as to the matter of fact, this echo is very wonderful." He continued:

> Such a musical canon might be contrived for one fine voice here . . . as would have all the effect of two, three, and even four voices. One blow of a hammer produced a very good imitation of an ingenious and practiced footman's knock at a London door, on a visiting night. A single ha! became a long horse-laugh; and a forced note, or a sound overblown in the trumpet, became the most ridiculous and laughable noise imaginable.[2]

The differences in approach of Kircher and Burney reflect in part a fundamental change in what the study of sound implied: for Kircher, music was a branch of mixed mathematics, and unsurprisingly his description is a geometric account of the felicitous proportions of the courtyard (see figure 1.1). Burney's investigation was less concerned with the principles of acoustics; rather, he was interested in the *experience* of echo, and his lively description allows readers to be virtual witnesses—or virtual auditors—of that experience (see figure 1.2).[3] Central to Burney's description is the notion of effect, here understood as phenomena that produced special or notable bodily reactions. This move from a mathematical to an experiential account is significant. Burney's description of the Villa Simonetta is indicative of a more general shift in the eighteenth century toward aesthetic modes of inquiry into music and—more generally—music's burgeoning autonomy as a discipline and object of knowledge.

Burney's motivations for visiting and describing the villa are not obvious. After all, it could be construed as not immediately useful to his conception of musical history and thus an indulgent digression. When in 1832 the English magazine the *Harmonicon* published a short biographical sketch of Burney's life, his trips were framed as doggedly and single-mindedly focused on one purpose:

> He resolved to make a tour of Italy and Germany, determined to hear with his own ears and see with his own eyes; and, if possible, to hear and see *nothing but music*. For although he might have amused himself enough in examining pictures, statues, and buildings, it was necessary he should economize the little time he could afford to be absent himself from England; and he could not indulge in general observation without neglecting the chief business of his journey.[4]

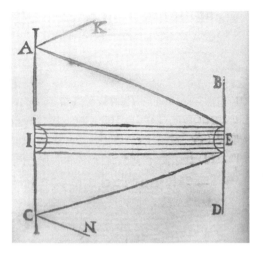

Figure 1.1. Image from Athanasius Kircher's *Musurgia Universalis*, book 9 ([Rome: Francisci Corbelletti, 1650], 290) demonstrating how sound bounces repeatedly from one wall to the wall opposite. (Jean Gray Hargrove Music Library, University of California, Berkeley.)

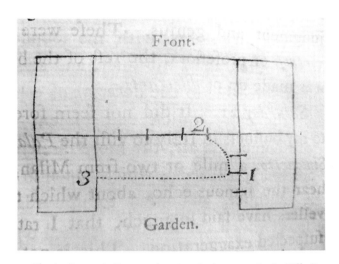

Figure 1.2. Charles Burney's diagram showing the key spots in the Villa Simonetta: "1. The best window to make the experiment at. 2. The best window to hear the echo. 3. A dead wall with only windows painted upon it, from whence the repetitions seem to proceed." (*The Present State of Music in France and Italy* [London: T. Becket, 1771], 99. Jean Gray Hargrove Music Library, University of California, Berkeley.)

One could argue that the villa formed part of the sonic landscape of Milan and that Burney's travels were far ranging and all-encompassing: he saw organs and military bands, street musicians and opera. In other words, Burney may indeed have focused on "nothing but music," but his definition of music was considerably broader than we might expect. However, a brief glance through either one of the travel diaries reveals the *Harmonicon*'s panegyric to be pure fantasy. Burney's investigation of the echo at the Villa Simonetta was one of many interests that lay beyond the musical: in the midst of trips to opera houses, churches, salons, and concert halls, he visited a number of laboratories and observatories of physicists and astronomers, describing what he witnessed with the same lively enthusiasm that he used on musical visits. Indeed, at the end of his second tour he wrote, "If my leisure and abilities would have sufficed for so extensive a plan, I should have been glad to make the journal of this tour, *the present state of arts and sciences, in general.*"[5]

In what follows, I argue that Burney's approach was radical—though in ways that are invisible now. This essay explores two ways in which a view of Burney's engagement with documenting science expands our understanding of the particular way in which he approached music. First I consider instruments, the ideas of experiment, and the ways Burney emphasized technical detail and performative effect; second, I turn to the idea of natural history, which was a "universal donor" to various facets of eighteenth-century culture.[6] My aim is to describe, on the basis of both Burney's travels and the evidence of his history of music, what sort of thing—what sort of object of knowledge—music was understood to be.

CHARLES BURNEY, ASTRONOMER

For anyone interested in the intersections between music and science, Burney is an obliging subject. He was an amateur astronomer: in October 1769, shortly before he embarked on his travels to collect materials for his history, he published *An Essay towards a History of the Principal Comets that have Appeared since the Year 1742*. Since the early eighteenth century, interest in matters astronomical had surged; in Burney's memoirs, the period is described as "a moment when astronomy was the nearly universal subject of discourse."[7] This expansion reflected the growth of popular science, but was also indicative of new technological developments such as the spread of refracting telescopes, which enabled more accurate celestial observation. Many new phenomena were thus unveiled: James Bradley discovered the nutation of the earth (the wobble caused by the moon's gravitational pull); in the 1730s, Pierre Maupertuis proved that the planet was oblate and not, as

Jacques Cassini had predicted, prolate (that is, the earth's polar axis is shorter than its equatorial axis). Other celestial phenomena were highly celebrated: the transits of Venus in 1761, for instance, were of enormous importance because they could be used to determine the longitude of particular locations.[8]

Comets were an especially exciting topic. In 1705, astronomer Edmond Halley had speculated that the famed comets of 1456, 1531, 1607, and 1682 (the last of which he observed himself) were in fact the same comet, and he predicted that it would return in 1758. In 1758, this arrival was awaited with great anticipation—and some trepidation, since comets were in this period still seen as signs of impending disaster. By Christmas Day of that year there had been no sighting, and all hope was almost lost; but then it appeared, causing a sensation. With the aid of better telescopes, new comets were being discovered yearly. Burney's ninety-three-page booklet was not, as his biographer Roger Lonsdale and others have suggested, motivated by the return of Halley's comet, but rather by the appearance of a new comet in 1769, one that the French astronomer Charles Messier had discovered.[9]

In 1789, after Burney completed his history of music, he again turned to the stars, beginning a poem on the history of astronomy. In 1797, while recovering from the death of his second wife Elizabeth the year before, his daughter Fanny encouraged him to return to the project and increase its scope. He imagined a lengthy project entitled *Astronomy, an historical & didactic Poem*, expected to fill twelve books. He shared his verses with his friend, the famed musician-astronomer William Herschel. Herschel offered both advice and support to Burney on his project:

> He gave me the greatest encouragement; said repeatedly that I perfectly understood what I was writing about; and only stopped me at two places. . . . The doctrine he allowed to be quite orthodox . . . he told me—that he had never been fond of Poetry—He thought it to consist of the arrangement of fine words, without meaning—but when it contained science and information, such as mine (hide my blushes!) he liked it very well.[10]

A fragment of the poem runs as follows:

> The various Orbs, pow'r & wisdom plann'd,
> Which float in Aether by divine command;
> The Laws immutable by wch they run
> In never-ending circles round the sun;
> The distance, magnitude, & motion's laws,
> Derived from Nature's Lord, the great First cause;

The Sages who with toil essay'd to find
The emanations of the Mighty Mind;
The means they've us'd in science to excel,
The Astronomic Muse essays to tell.[11]

Burney never completed his "historical & didactic poem"—he ended up committing eight completed volumes to the flames, finding that not all of his readers were as enthusiastic as Herschel. In 1965, Lonsdale rejoiced in what he perceived to be their mediocrity, setting the trend for the subsequent reception of Burney's poetic works. H. C. Robbins Landon, for example, felt no hesitation in slighting them, writing of Burney's *Verses in praise of Haydn's arrival in England* that "Dr. Burney was many things: a good 'scholastic' composer, a brilliant historian, and a man of much taste, learning, and wit. A poet he was not."[12] However, we should not too hastily dismiss Burney's poetic output: his verses on Haydn, and his astronomical poem, sought to mix education and pleasure, to delight as well as to inform. In this, they partook of a venerable tradition of didactic poetry that stretched back to antiquity.

Jesuit scholars, for example, often published poetic works alongside their dissertations. Both forms were a means of disseminating knowledge. As Yasmin Haskell has argued, the reader of such verses has to invest emotionally in order understand them, thus participating actively in discovery.[13] Such work exercises the reader's intellect while simultaneously appealing to sentiment: the feeling of discovery is an *effect* that is created by carefully constructed verse. Didactic poetry also offered the possibility to bring its subject to life in ways that could reach a wide audience. The naturalist Erasmus Darwin (1731–1802), Burney's contemporary, helped to introduce the Linnaean system of classification of plants in Britain. After translating Linnaeus's works into English between 1783 and 1787, he anonymously published an extended poem, *The Loves of the Plants*, in 1789. A deliciously sensual and playful introduction to botany and an extremely successful publication, it helped popularize the notion that plants had gender.[14] Many scholars stress how Darwin anthropomorphized plants within the poem, though he had already used the language of husbands, wives, marriage, beds, and homes in his earlier, unversed translations. What is striking in this poem is the *animation* of plant life; to understand the notion that plants had sex lives, they first had to have lives: flowers acted, seduced and loved ("What Beaux and Beauties crowd the gaudy groves / And woo and win their vegetable Loves").[15] As Mark Lussier has noted, Darwin's poetry exerted influence over poets during the 1790s and beyond, even as his natural-historical work became less popular:

the conception of nature that appears in William Blake's and Samuel Taylor Coleridge's poems owes much to Darwin's verses.[16] "To conduct to any science by a path strewed over with flowers," Fanny wrote to her father in 1797, "is giving beauty to labor, and making study a luxury."[17]

Darwin's amorous flowers and Burney's celestial couplets point to the larger ways in which the production of knowledge and feeling were imbricated in this period. Burney was a privileged observer of culture, and his interest in matters of astronomy and natural philosophy both preceded and followed the publication of his major musical writings. For the purposes of this volume, the ways in which his lifelong engagement with science was closely bound up with his activity as a scholar of music may serve as a useful reminder. To think about music and science is not necessarily to confuse two separate cultures or to reveal secret connections between disparate realms. Burney's activities point to deep, epistemological connections between a range of fields—astronomical, musical, and natural-historical—in the late eighteenth century.

WHAT BURNEY SAW

In order to understand Burney's particular approach to writing and thinking about music and how it was imbricated with notions of experiment, it is useful to look closely at the kinds of observations he made on his tours. Burney pioneered a particular approach to describing his encounters with instruments and their performance. When in Rome, for example, he went to the Basilica of St. John Lateran, where he saw the largest organ in the city:

> I was conducted into the great organ-loft by Signor Colista, who did me the favour to open the case, and to shew me all the internal construction of this famous instrument. . . . It has thirty-six stops, two set of keys, long eighths, an octave below double F. and goes up to E. in *altissimo*. It has likewise pedals; in the use of which Signor Colista is very dextrous. His manner of playing this instrument seems to be the true organ stile, though his taste is rather ancient. . . . Signor Colista played several fugues, in which the subjects were frequently introduced on the pedals, in a very masterly manner.[18]

His rhetoric intensified when the subject moved from more familiar musical instruments to less familiar ones. When in Milan, he visited the Jesuit scholar Roger Boscovich, an astronomer, physicist, mathematician, and philosopher who published on a wide range of subjects, from the construction of telescopes to ancient sundials to the aurora borealis. The Jesuit also published poetry, including a six-book poem on eclipses. Burney reported that Boscovich received him enthusiastically: "Being told that I was an Englishman, a lover of the sciences, and ambitious of seeing so celebrated a man, he addressed himself to me in a particular manner." Burney continued:

> He immediately began to shew and explain to me the construction and use of several machines and contrivances which he had invented for making optical experiments. . . . He then went on, and surprised and delighted us all very much, particularly with his *Stet Sol*, by which he can fix the sun's rays, passing through an aperture or a prism, to any part of the opposite wall he pleases: he likewise separates and fixes any of the prismatic colors of the rays. Shewed us a method of forming an aquatic prism, and the effects of joining different lenses, all extremely plain and ingenious. Then we ascended to different observatories, where I found his instruments mounted in so ingenious and so convenient a manner, as to give me the utmost pleasure.[19]

The *Stet Sol* was probably a scioptic ball—a moveable lens modeled on the human eye that could be used to focus the sun's rays, either with telescopic or microscopic lenses. The aquatic prism was an adjustable V-shaped vessel that could be filled with liquids of varying densities for optical experiments. While Burney's description of the St. John Lateran organ was relatively straightforward, here, when describing less familiar objects, his language brims with enthusiasm.

Later on his travels, in Bologna, Burney met the physicist Laura Bassi, who had defended her doctorate in 1732, becoming only the second woman in Europe to receive a doctoral degree. She was one of the first scholars to teach Newtonian natural philosophy in Italy.[20] Like Boscovich, she wrote poems; she also made appearances at public events, and was much celebrated in her own day, often depicted as Minerva. Burney was delighted to meet her: "This lady is between fifty and sixty; but though learned, and a genius, not at all masculine or assuming." What was more,

> She shewed me her electrical machine and apparatus: the machine is simple, portable, and convenient; it consists of a plain plate of glass, placed vertically; the two cushions are covered with red leather; the receiver is a tin forked tube; the two forks, with pins at the ends, are placed

next [to] the glass plate. She is very dexterous and ingenious in her experiments, of which she was so obliging to shew me several.[21]

Bassi's "electrical machine" was a generator that used friction to create an electrostatic charge. Such devices became increasingly elaborate in the mid-eighteenth century and were used in both experiments and spectacular displays.[22]

Some of the most memorable moments in Burney's accounts are those in which his own awe in the face of firsthand experience becomes palpable. During his second journey, while in Hamburg, Burney had the pleasure of hearing C. P. E. Bach perform on his Silbermann clavichord, which he did with "delicacy, precision, and spirit":

> I prevailed upon him to sit down again to a clavichord, and he played, with little intermission, till near eleven o'clock at night. . . . During this time, he grew so animated and *possessed*, that he not only played, but looked like one inspired. His eyes were fixed, his under lip fell, and drops of effervescence distilled from his countenance. . . . His performance to-day convinced me of what I had suggested before from his works; that he is not only one of the greatest composers that ever existed, for keyed instruments, but the best player, in point of *expression*.[23]

When Burney was in Ludwigsburg, he saw a grand orrery by Philipp Matthäus Hahn, the "priest mechanic" who was a clockmaker and the creator of the first mechanical calculator. Hahn's orrery was particularly spectacular: it showed the time—day, month, and year—and the motion of the planets and their relation to the major constellations. As in the case of Bach's performance, the instrument induced in Burney a feeling of awe:

> This whole machine is so constructed, that without any risk of putting it out of order, or spoiling it, the reciprocal position of the planets and constellations, such as they *will* be in any future minute, or such as they *have* been, in any one that is past, may be seen, so that this machine takes in all time; the past, the present, the future; and is, not only an orrery for these times, but a perpetual, accurate, and minute history of the heavens for all ages.[24]

As beautiful objects, orreries could enchant and fascinate as much as instruct; Joseph Wright drew upon this power in his famous painting of the philosopher's lecture on the orrery. In all of these descriptions, Burney emphasizes the effect of witnessing the various instruments and their demonstration. Indeed, his enraptured account of their effects seems to stand in for a description of the actual experiments carried out or the music performed.

Two aspects of these descriptions are of particular interest here: the first is Burney's attention to the various instrumental and technical configurations he encountered. The second concerns the feelings of awe, delight, astonishment, and admiration that they produced for Burney the observer. These "feelings," along with one's susceptibility to them, in other words, played a vital role in his conception of perspicacious scientific activity. Given his pursuit of instrumental effects in all their vivid variety, his writing can thus productively be seen as emerging from more general trends in natural philosophy, as in the "rise of the experimental method" with all its concomitant assumptions about knowledge production and instrumental precision.

Exactly what constituted the experimental method and how it emerged is a complicated topic, whose exploration lies outside the scope of this chapter. It is, however, worthwhile to alight on one aspect: the notion that the act of witnessing experimental demonstrations produced new forms of knowledge, which could then be disseminated and explained. This idea, rather than being universal to the history of science, is one that emerged in the late seventeenth century. As Simon Schaffer and Steve Shapin have argued in their now-classic book *Leviathan and the Air Pump*, what was radical about Robert Boyle's experiments with the air pump was the notion that experiment and demonstration actually could create "matters of fact."[25] For Boyle and many others within the nascent Royal Society, demonstration and experiment revealed truths of nature, and, in the process of revealing those truths, gave rise to consensus among the spectators. The air pump, as Schaffer and Shapin show, then became a form of literary technology when experiments were disseminated in written form, allowing the possibility of virtual witnessing. In similar fashion, one might argue, Burney's literary style made virtual auditors of his readers, in that they were persuaded to consensual experiences and emotions.

The turn toward experiment also marked a move away from "pure reason" and toward the senses, empirical research, and instruments. Within this broader context of his scientific epistemology, Burney's descriptions shed light on the connection between the rise of experimentation and the heighted attention to the nature of and roles played by the instruments and technologies that aided and extended the senses. This focus both reflected the central position of instruments within experiment, and also pointed to the ways in which this new, central role necessitated a closer understanding of their function. What we see through a microscope, for example, can only give rise to new matters of fact if we trust that the microscope is magnifying something that existed in reality before we peered down the eyepiece; the instrument only works when we believe that it is not creating new phenomena. In other words, experimentation brings focus not just to instruments but also to modes of sensory mediation.

This notion of the mediator can also be applied to Burney himself. His awareness of the effects produced by the sonic phenomena he experienced drew attention to his subjective role as a feeling observer. Jonathan Crary's seminal *Techniques of the Observer* deals at length with problems of bodily mediation by emphasizing an important shift in the understanding of vision in the early nineteenth century, epitomized by Goethe's *Zur Farbenlehre* (1810).[26] Goethe famously critiqued Newton's theory of light, showing the centrality of "psychological colors" to our perception: that is, colors were not simply out in the world to be passively perceived by observers. Rather, our eyes actively produce colors. The phenomena of afterimages and images produced by gently pressing on the side of the eye, for example, were proofs that colors, in Goethe's words, "belong to the eye."[27] More recently, Jonathan Sterne has explored an auditory analog to Crary's argument. In *The Audible Past*, Sterne speaks of "ensoniment"—the process, beginning in the eighteenth century, through which sound itself became an "object and domain of thought and practice" and was thereby "reconstructed as a physiological process." Sterne casts this psychological turn as the birth of a new discourse of sound and hearing different from the more "idealized" conception of music, to use Sterne's language—the assumption being that late eighteenth- and early nineteenth-century composers, musicians, and writers on music could carry on thinking of music *out there* without necessary probing the nature of the organs of hearing. A careful overview of Burney's writings, however, forces us to give nuance to any such blunt articulation of overly stark divides between music and other sonic phenomena. To be sure, Burney's enthusiastic descriptions of instruments, sonic phenomena, performers, and their effects are worlds away from the mid-nineteenth-century German physiologist Johann Müller's investigations into the auditory nerve (to cite one of Sterne's important case studies).[28] Nevertheless, there can be little doubt that Burney's emphasis on and keen attention to aural effect and instrumentation did prefigure and form part of the preconditions to later physiological studies of the ear.

Placing Burney's writing into this broader history of mediation also sheds light on the connection between the rise of experimentation and what recent writers have called "the birth of modern aesthetics." Aesthetics, in its original Baumgartian sense, was "a science of how things are to be known by means of the senses."[29] That is, it was an inquiry into the process by which our sensations of the outside world are translated into higher orders of thought, and a study of how our senses present the outside world for our interior world of ideas. Aesthetics began as a category of empirical episte-mology, which came into being, as Michael McKeon aptly puts it, through "explicit emulation of a normative model of . . . scientific cognition."[30] Aes-thetics, when seen in this light—and not simply as a study of the beautiful— is first a study of mediation and is a mediating force itself, for it deals in equal measure with immediate sensation and abstract reason. Just as the experi-mental method necessitated a closer understanding of the instruments that aided the senses, the emergence of a discourse on aesthetics likewise invited attention to the instruments and technologies that made aesthetic experi-ence possible. Within musical culture, such attentiveness to instrumentality took a variety of forms: as I have argued elsewhere, new discourses on the subject of "timbre" made new forms of aesthetic attention possible, as did the concomitant changes to practices of orchestration.[31] Outside of compo-sitional practice, aesthetic preoccupations also explain why understanding music's technological configurations was deemed so essential to communi-cation about musical experience.

In this respect, Burney can be usefully characterized as an experimental philosopher. He was certainly engaged in experimental pursuits: he made careful, firsthand observations about musical culture, which he then had to communicate to his readers.[32] Wherever he went, he systematically cata-logued and described both the musical instruments he encountered and his experiences of them: the brilliant orchestras, the powerful military bands, the out-of-tune organs, and the clanging carillons. To form an accurate por-trait of the musical life of a city, Burney took it upon himself to first under-stand the instruments that were available to that city and then communicate the effect of those instruments to his readers. Burney's quest for swells on German organs, his concerns over intonation, his impatience with musical clocks, even his bafflement at the vocal style in the synagogue in Amster-dam—which sounded to him like a strange imitation of flutes, bassoons, and violins—can all be seen as part of his intimate attention to instrumentation.

Sensitivity toward instruments and the experiences they afforded not only marked Burney out as a modern historian. Sensitivity was also the crite-ria by which he judged composers and performers: he sought out and prized sensitive and nuanced handling of instruments, and complained when they were mishandled. While he was Berlin, for example, he grumbled about a lack of instrumental feeling and dynamic contrast in German music:

> If I may depend upon my own sensations, I should imagine that the musical performances of this country want *contrast*; and there seems to be not only too many notes in them, but those notes are expressed with too little attention to the *degree* of force, that the instruments, for which they are made are capable of. Sound can only be augmented to a certain degree, beyond that, is *noise*.[33]

What Burney made consistently clear in his observations was that good music required not only sensitive musicians but also compositions that displayed a true understanding of instruments, their acoustical limitations, and their expressive capacity. Throughout his travelogues, Burney both emphasized and produced a particular kind of knowledge about the musical cultures of the cities he visited, one that was formed in the act of witnessing, and then communicated to his readers through his lively and enthusiastic descriptions. Burney's clear enthusiasm for his subject is not an adornment to his travel reports. As virtual witnesses to his travels, his readers were afforded the chance to *feel*, and that feeling helped them to join in verifying the sensible truth of his observations. Perhaps we can think of Burney himself as a kind of mediating instrument: he was a sensitive being that collected empirical data and then disseminated it in literary form for his readers to consume.

TOWARD A NATURAL HISTORY OF MUSIC

Understanding the role of careful observation in Burney's travels ultimately allows us to return to the query posed at the beginning of this chapter: what kind of object was music? In the introduction to the first volume of Burney's history, he offered the following disclaimer:

> With respect to the present work, there may, perhaps, be many readers who wish and expect to find in it a deep and well digested treatise on the theory and practice of music: whilst others, less eager after such information, will be seeking for mere amusement in the narrative. I wish it had been in my plan and power fully to satisfy either party; but a history is neither a body of laws, nor a novel. I have blended together theory and practice, facts and explanations, incidents, causes, consequences, conjectures, and confessions of ignorance, just as the subject produced them.[34]

Burney's approach to his history set it apart from other histories of music. Some of the difference lay in the style of the work: the first volume coincided with the publication of Sir John Hawkins's *General History of the Science and Practice of Music*. The two authors were unsurprisingly seen as musical rivals and readers found Burney's history more charming and lively, leading, famously, to the witty catch by John Wall Callcott, which began "Have you Sir John's Histr'y / Some folks think it quite a myst'ry." It ended with repeated cries of "Burney's hist'ry pleases me," which sounded in performance like a rallying cry: "Burn his history!" But the stylistic difference also pointed to a more fundamental distinction between their research methods: Hawkins, unlike Burney, did not base his history on first-hand observation but rather on research gleaned from books and manuscripts. Burney told the story of music as a vibrant tradition whose development revealed progress, leading to the glorious present. Hawkins, in contrast, celebrated glories of music's past. Likewise, Padre Martini's *Storia della musica*, published between 1757 and 1781, placed emphasis on the science of harmony, which was shown to take on different manifestations in different epochs of musical history. In other words, for Martini music itself didn't evolve: it just wore different fashions.[35] Burney's history also differed from Giorgio Vasari's *Lives of the Most Excellent Painters, Sculptors and Architects*, which, as the title suggests, focused first and foremost on individual artists.[36] Certainly, Burney's history occasionally adopted an explicitly biographical focus, but it was not his dominant organizational method. The question is: if Burney did not appeal to a static or mathematical theory as unifying principle, what then is the organizational principle informing his history?

In Vanessa Agnew's imaginative study of Burney, she draws useful connections between him and Captain James Cook: both men, she argues, were explorers; Cook's first great voyage coincided with Burney's first tour.[37] As enticing as this parallel is, I would like to tweak it slightly. Burney did not navigate uncharted lands. Rather, his own practices during his travels more closely resembled those of Joseph Banks, the natural historian who accompanied Cook on the first great voyage and who collected, catalogued, and classified the new species he encountered. Banks was in fact a correspondent and friend of Burney, the latter having addressed his fellow naturalist in a 1791 letter as the "Patron and Friend to all deserving circumnavigators."[38] Like Banks, Burney was a collector, only mostly of music—from the past and as it existed in the present—which he classified according to the diverse genres, instrumentalities, national types, and "species" he encountered.

In 1774, Burney sent a series of queries to Canton, China, addressed to an Italian missionary. These letters asked for information about the scales used, whether melodies ever modulated and whether harmony was cultivated; Burney also asked for a "specimen of the *Pierres sonores*" and a general list of the important musical instruments, along with "a few tunes or compositions for each."[39] The answers came back too late to be included in the first edition of his *General History*; he amended the second edition to include the information he had gleaned from the "specimens that [he] was able to collect."[40] Many of the questions reflected Burney's particular interest in the similarities between Scottish and Chinese tunes. The queries, however, reveal something about how he conceptualized the objects he was studying: his specimens could be melodic fragments or instruments. This interchangcability of instrument and notated music is reflected elsewhere in his historical project: where Burney couldn't access or experience actual musical performances—for example, with the music of the ancient world—he collected and relied on instruments, music's material traces.[41] For Burney, it seemed, music as a general idea borrowed the object status of instruments. His collection and classification of these various objects of musical culture, in other words, had profound significance for what music was, as an object of knowledge.

We can better understand the importance of this natural-historical turn for questions of musical epistemology by returning to astronomy and its radical transformation in the late eighteenth century. Astronomy, like music, had traditionally been considered a branch of mathematics. This changed in the hands of Burney's aforementioned friend, William Herschel. Herschel was the discoverer of the infrared part of the spectrum and of Uranus—the first planet to be discovered with a telescope. Like Burney, as a young man he made a living as a musician; the first part of his life was spent not among telescopes and micrometers, but with oboes, violins, and keyboard instruments. From 1753 to 1756, he held a post as oboist and violinist in the Hanover Guards. In 1756, he moved to England, where he had various jobs largely outside London, performing and composing music. In late 1766, he was hired as organist at the Octagon Chapel in Bath, which is to say that his turn to astronomy happened gradually. In the 1770s, he began to construct his own telescopes and stands, meticulously grinding his own lenses.[42] Later, having been appointed the court astronomer of King George III, he built an impressive forty-foot telescope in his garden at Slough, which Joseph Haydn famously saw on one of his visits to England. His interaction with his instruments seems to have been shaped by notions from music. In a letter describing discoveries that he had made with a new telescope, he wrote:

I do not suppose there are many persons who could ever find a star with
my power of 6,450, much less keep it, if they found it. Seeing is in some
respects an art, which must be learnt. To make a person see with such a
power is nearly the same as if I were asked to make him play one of Handel's fugues on the organ. Many a night have I been practicing to see, and
it would be strange if one did not acquire a certain dexterity by such constant practice.[43]

Schaffer has argued that what was most transformative about Herschel's
project is that his delicate and careful observations, aided by his precision
instruments, allowed him to *classify* the heavens. Having "practiced to see"
and achieved "a certain dexterity" on these instruments, in other words, Herschel was able to identify a set of natural-historical objects. Schaffer writes,
"By isolating species among these specimens and arranging these species in
sets of connected series, Herschel as a true natural historian reconstituted
the natural order of the heavens."[44] What was more, Herschel's cosmology
was a living one—making it possible to understand the ways in which the
cosmic bodies were neither absolute nor stable, but instead changed over
time; his observations revealed that the stars themselves had a history. It was
precisely this conception that made possible the nebular hypothesis, that is,
the notion that the heavens grew out of a state of chaos.

For Burney, music was likewise an object of natural history. Through the
painstaking inspection and audition of instruments from across the world,
he developed a classificatory system that was not mathematical but rather
based on sensible, observable attributes of its objects. He explicitly equated
music with the natural world: "Music, indeed, like vegetation, flourishes differently in different climates; and in proportion to the culture and encouragement it receives."[45] Toward the end of the fourth and final volume of
his history, he offered the following recommendation: "I would advise true
lovers to Music to *listen* more than talk, and give way to their feelings, not
lose the pleasure which melody, harmony, and expression ought to give, in
idle inquiries into the nature and accuracy of their auricular sensations."[46]
For a reader coming to the end of nearly 2,500 pages of musical history, this
statement might appear strange, even paradoxical. But this is precisely how
Burney went about writing his history: by embracing music's pleasures, he
was able to construct a history that navigated a special middle ground. His
history was not a history of the laws of music; nor did it delve into great
biographical detail. But between abstract theory and personal narrative lay
music, music as practiced and performed throughout history and across the

European continent, in all its many styles and forms—in all of its species. Music was an independent, classifiable object, one that changed and evolved over time. It was, like the starry canopy, teeming with energy and life.

NOTES

1. See Athanasius Kircher, "De Mirifica Echo Villae Simonetta Mediolani," in *Musurgia Universalis, Sive Ars Magna Consoni et Dissoni in X. Libros Digesta* (Rome: Corbelletti, 1650), vol. 9, part 4, 289–91. On the echo and its history, see Iris Lauterbach, "The Gardens of the Milanese 'Villeggiatura' in the Mid-Sixteenth Century," in *The Italian Garden: Art, Design and Culture*, ed. John Dixon Hunt (Cambridge: Cambridge University Press, 1996), 127–59.

2. Charles Burney, *The Present State of Music in France and Italy: or, The Journal of a Tour through those countries, undertaken to collect Materials for A General History of Music*, 2nd ed. (London: Becket, 1773), 104. The "footman's knock" was of a special kind. *Blackwood's Edinburgh Magazine* provides the following knock taxonomy in 1824:

> 1. Hypocrousis.—A modest timid inaudible knock. 2. Monocrousis—The plain single knock of a tradesman *coming for orders*. 3. Dicrousis.—The postman and taxgatherer. 4. Tricrousis.—The attempt of the same tradesman to express hi [*sic*] impatience, and compel payment of his bill; he will not submit to the *single* knock any longer, and dare not venture on *the following*. 5. Tetracrousis—Your own knock; my own knock; a gentleman's knock. 6. Pollacrousis, or *Keraunos*.—A succession of repeated impulses of different tone and force, ending in three or four of *alarming emphasis*—vulgo, a footman's knock, a thundering knock., &c. &c. &c. ("Once More in London," *Blackwood's Edinburgh Magazine*, January 1824, 97–98).

3. Steven Shapin, "Pump and Circumstance: Robert Boyle's Literary Technology," *Social Studies of Science* 14, no. 4 (November 1984): 481–520.

4. "Memoirs of Charles Burney," *Harmonicon* 10 (1832): 215 (original emphasis).

5. Charles Burney, *The Present State of Music in Germany, the Netherlands, and United Provinces* (London: Becket, Robson, and Robinson, 1775), 2:332 (original emphasis).

6. See, e.g., Emma Spary, "Political, Natural, and Bodily Economies," in *Cultures of Natural History*, ed. N. Jardine, J. A. Secord and E. C. Spary (Cambridge: Cambridge University Press, 1996), 178–96.

7. D'Arblay and Burney, *Memoirs of Doctor Burney*, 1:219.

8. Elaine Sisman has explored these transits in relation to Joseph Haydn's preoccupation with the sun in 1761; see Sisman, "Haydn's Solar Poetics: The Tageszeiten Symphonies and Enlightenment Knowledge," *Journal of the American Musicological Society* 66, no. 1 (April 2013): 5–102.

9. Burney's interest in astronomy was something he had shared with Esther (1725–1762), his first wife. She translated Maupertuis's 1742 "Letter Upon Comets," which was included at the beginning of Burney's essay; see Mary Terrall, *The Man Who Flattened the Earth: Maupertuis and the Sciences in the Enlightenment* (Chicago: University of Chicago Press, 2002).

10. Charles Burney in a letter to Fanny Burney (September 28, 1797), published in *Diaries and Letters of Madame d'Arblay (1788–1840)*, vol. 5, ed. Charlotte Barret (London: MacMillan, 1905), 346.

11. From a surviving fragment of about four hundred lines in the Burney Family Collection OSB MSS 3 (box 5, folder 332, Beinecke Rare Book and Manuscript Library, Yale University, New Haven, CT); quoted in Roger H. Lonsdale, *Dr. Charles Burney: A Literary Biography* (Oxford: Clarendon Press, 1965), 404.

12. H. C. Robbins Landon, *Haydn: Chronicle and Works*, vol. 3 (London: Thames and Hudson, 1994), 32.

13. Yasmin Haskell, *Loyola's Bees: Ideology and Industry in Jesuit Latin Didactic Poetry* (Oxford: Oxford University Press, 2003).

14. On the unexpected popularity of Darwin's verses, see Janet Browne, "Botany for Gentlemen: Erasmus Darwin and 'The Loves of the Plants,'" *Isis* 80, no. 4 (1989): 592–621. On Darwin's influence, see Patricia Fara, *Erasmus Darwin: Sex, Science, and Serendipity* (Oxford: Oxford University Press, 2012).

15. Erasmus Darwin, *The Botanic Garden: A Poem, in Two Parts. Part II. The Loves of the Plants. With Philosophical Notes* (London: J. Johnson 1791), 2.

16. Mark Lussier, "Science and Poetry," in *The Encyclopedia of Romantic Literature*, ed. Frederick Burwick, Nancy Moore Goslee, and Diane Long Hoeveler (Malden, MA: Wiley-Blackwell, 2012), 3:1184.

17. Quoted in Lonsdale, *Dr. Charles Burney*, 386.

18. Burney, *The Present State of Music in France and Italy*, 387.

19. Ibid., 86–87.

20. On Bassi, see Paula Findlen, "Science as a Career in Enlightenment Italy: the Strategies of Laura Bassi," *Isis* 84, no. 3 (1993): 441–69; Findlen, "A Forgotten Newtonian: Women and Science in the Italian Provinces," *The Sciences in Enlightened Europe*, ed. William Clark, Jan Golinski, and Simon Schaffer (Chicago: University of Chicago Press, 1999), 313–49; and Alberto Elena, "'In lode della filosofessa di Bologna': An Introduction to Laura Bassi," *Isis* 82, no. 3 (1991): 510–18.

21. Burney, *The Present State of Music in France and Italy*, 218.

22. See, e.g., James Delbourgo, *A Most Amazing Scene of Wonders: Electricity and Enlightenment in Early America* (Cambridge, MA: Harvard University Press, 2006), esp. ch. 3, "Wonderful Recreations."

23. Burney, *The Present State of Music in Germany, the Netherlands, and United Provinces*, 2:269–70 (original emphasis).

24. Burney, *The Present State of Music in Germany, the Netherlands, and United Provinces*, 1:118 (original emphasis).

25. Steven Shapin and Simon Schaffer, *Leviathan and the Air-Pump: Hobbes, Boyle, and the Experimental Life* (Princeton, NJ: Princeton University Press, 1985).

26. Jonathan Crary, *Techniques of the Observer: On Vision and Modernity in the Nineteenth Century* (Cambridge, MA: MIT Press, 1990).

27. Jonathan Crary, "Techniques of the Observer," ch. 4 in *Techniques of the Observer*, 104ff. Crary likewise shows how this new form of attention also had its own particular technological implications. The notion of afterimages soon gave rise to a barrage of devices—including the thaumatrope, phenakistoscope, and zellotrope—that used the properties of perception to create precinematic optical effects.

28. Jonathan Sterne, *The Audible Past: Cultural Origins of Sound Reproduction* (Durham, NC: Duke University Press, 2003); see in particular his discussion of Johann Müller on pp. 60–64.

29. Alexander Gottlieb Baumgarten, *Meditationes philosophicae de nonnullis ad poema pertinentibus* (Halle: J. H. Grunerti, 1735), cxv–cxvi.

30. Michael McKeon, "Mediation as Primal Word: The Arts, the Sciences, and the Origins of the Aesthetic," in *This Is Enlightenment*, ed. Clifford Siskin and William Warner (Chicago: University of Chicago Press, 2010), 385.

31. Emily I. Dolan, *The Orchestral Revolution: Haydn and the Technologies of Timbre* (Cambridge: Cambridge University Press, 2013); see esp. ch. 2, "The Birth of Timbre."

32. On the role of sentimentality in scientific inquiry, see also Jessica Riskin, *Science in the Age of Sensibility: The Sentimental Empiricists of the French Enlightenment* (Chicago: University of Chicago Press, 2002).

33. Burney, *The Present State of Music in Germany, the Netherlands, and United Provinces*, 2:202–3 (original emphasis).

34. Burney, *General History of Music*, 1:xvii. He continues:

> Many new materials concerning the art of Music in the remote times of which this volume treats, can hardly be expected. The collecting into one point the most interesting circumstances relative to its practice and professors; its connection with religion; with war; with the stage; with public festivals, and private amusements, have principally employed me: and as the historian of a great and powerful empire marks its limits and resources; its acquisitions and losses; its enemies and allies; I have endeavored to point out the boundaries of music, and its influence on our passions; its early subservience to poetry, its setting up a separate interest, and afterward aiming at independence; the heroes who have fought its battles, and the victories they have obtained.

35. See, e.g., Ivano Cavallini, "L'idée d'histoire et d'harmonie du Padre Martini et d'autres penseurs de son temps," *International Review of the Aesthetics and Sociology of Music* 21, no. 2 (1990): 141–59.

36. Vasari, *La vita de' più eccellenti pittori, scultori ed architettori* (1550; rev. ed., Florence: Giunti, 1568).

37. Vanessa Agnew, *Enlightenment Orpheus: The Power of Music in Other Worlds* (Oxford: Oxford University Press, 2008).

38. Burney to Banks, July 6, 1791; quoted in Neil Chambers, "Letters from the President: The Correspondence of Sir Joseph Banks," *Notes & Records of the Royal Society of London* 53, no. 1 (1999): 41.

39. "Musical Queries for China" [1774], Burney Family Collection, OSB MSS 3, box 5, folder 354, Beineke Rare Book and Manuscript Library, Yale University, New Haven, CT.

40. Burney, *General History of Music*, 1:31.

41. Zdravko Blažeković discusses the history, significance of, and visual practices behind the engravings of instruments in Burney's history; see "Vesuvian Organology in Charles Burney's *General History of Music*," in *Klänge Der Vergangenheit: Die Interpretation von Musikarchäologischen Artefakten Im Kontext*, ed. Ricardo Eichmann, Fang Jianjun, and Lars-Christian Koch, Studien zur Musikarchäologie 8; Orient-Archäologie 27 (Rahden, Westf.: M. Leidorf, 2012), 39–57. The plates that accompany the first volume

include collections of instruments from the ancient world; what is notable here is that
these depictions are not generalized examples but concrete specimens that had been exca-
vated and particular ancient depictions and representations.

42. On Herschel, see Michael A. Hoskin, *Discoverers of the Universe: William and
Caroline Herschel* (Princeton, NJ: Princeton University Press, 2011).

43. Herschel to Watson (1782), in Constance A. Lubbock, *The Herschel Chronicle:
The Life-Story of William Herschel and his Sister, Caroline Herschel* (New York: Macmillan, and Cambridge: The University Press, 1933), 99–101.

44. Simon Schaffer, "Herschel in Bedlam: Natural History and Stellar Astronomy,"
British Journal for the History of Science 13, no. 3 (1980): 213.

45. Burney, *General History of Music*, 1:i.

46. Ibid., 4: 950, 973 (original emphasis).

CHAPTER TWO

Celestial Mechanisms:
Adam Walker's Eidouranion, Celestina,
and the Advancement of Knowledge

DEIRDRE LOUGHRIDGE

E dward Francis Burney's illustration of the English Opera House as it appeared on the evening of March 21, 1817, presents a fashionable audience before a grand spectacle (see figure 2.1). But the spectators had gathered to hear not the latest opera but rather an astronomical lecture on the eidouranion. Invented by Adam Walker in 1781, the eidouranion—or "grand transparent orrery"—was a large model solar system designed for display on the stage of a theater.[1] Also present, though not visible in Burney's illustration, was another of Walker's inventions: a harpsichord equipped with a mechanism for bowing the strings. This musical instrument he called a celestina.

As the inventor of the eidouranion and the celestina, and as one who deployed both instruments in popular lectures, Walker connected astronomy and music in concrete ways. But Walker was not much of a musician (apart from a bit of country fiddling); nor did he strive to make new astronomical discoveries. To most, he was a "lecturer in experimental philosophy"—one of the numerous eighteenth-century men who made Newtonian science accessible to the general public by eliminating its complicated mathematics and focusing on what could be demonstrated with instrumental apparatus. The caricaturist James Gillray captured Walker in this role, depicting him in the lecture room maintained at his home on Conduit Street, Hanover Square, London (see figure 2.2). Walker was also what Charles Burney (the music historian who purchased a celestina from him on behalf of Thomas Jefferson) called a "projector"—someone constantly planning new undertakings and thus "having too many pursuits at a time."[2] While most eighteenth-century lecturers in experimental philosophy combined lecturing with scientific instrument making and publication of educational treatises, Walker went beyond these occupational activities with his ventures in theatrical-scale performance and musical instrument making.

Figure 2.1. Edward Francis Burney, "The Proscenium of the English Opera House in the Strand (late Lyceum) as it appeared on the evening of the 21 March 1817, with [Dean] Walker's exhibition of the Eidouranion," as reproduced in Robert Wilkinson, *Londina illustrata*, vol. 2 (London: R. Wilkinson, 1819–25), n.p. (Doe Library, University of California, Berkeley.)

The pairing of music and astronomy seems a natural one, thanks to the long history of musical models of planetary motions based on a "harmony of the spheres." But Walker was one of a modern breed of experimental philosophers who actively rejected such speculative systems for understanding the universe, preferring instead that which could be demonstrated by observation with the help of instruments. Walker's efforts thus illustrate the conditions—social, aesthetic, commercial, epistemological—under which music and astronomy could form a new partnership. Walker called his eidouranion and celestina "sister inventions," suggesting they were born of the same impulse to demonstrate astronomy to large audiences (see figure 2.3). In fact, however, he had invented the celestina a decade earlier and under rather different circumstances. In the 1770s, Walker used the instrument to "enliven" a series of intimate lecture courses, which ranged over a wide variety of topics in natural philosophy. Following the path of Walker and his celestina from these earlier lectures to the grand spectacle of the eidouranion sheds light not only on the opportunities and values that guided Walker's work but also on the polyvalent function of musical instruments in relation to experimen-

Figure 2.2. James Gillray, *A Philosopher—Conduit Street*, depicting Adam Walker (1796). (© National Portrait Gallery, London.)

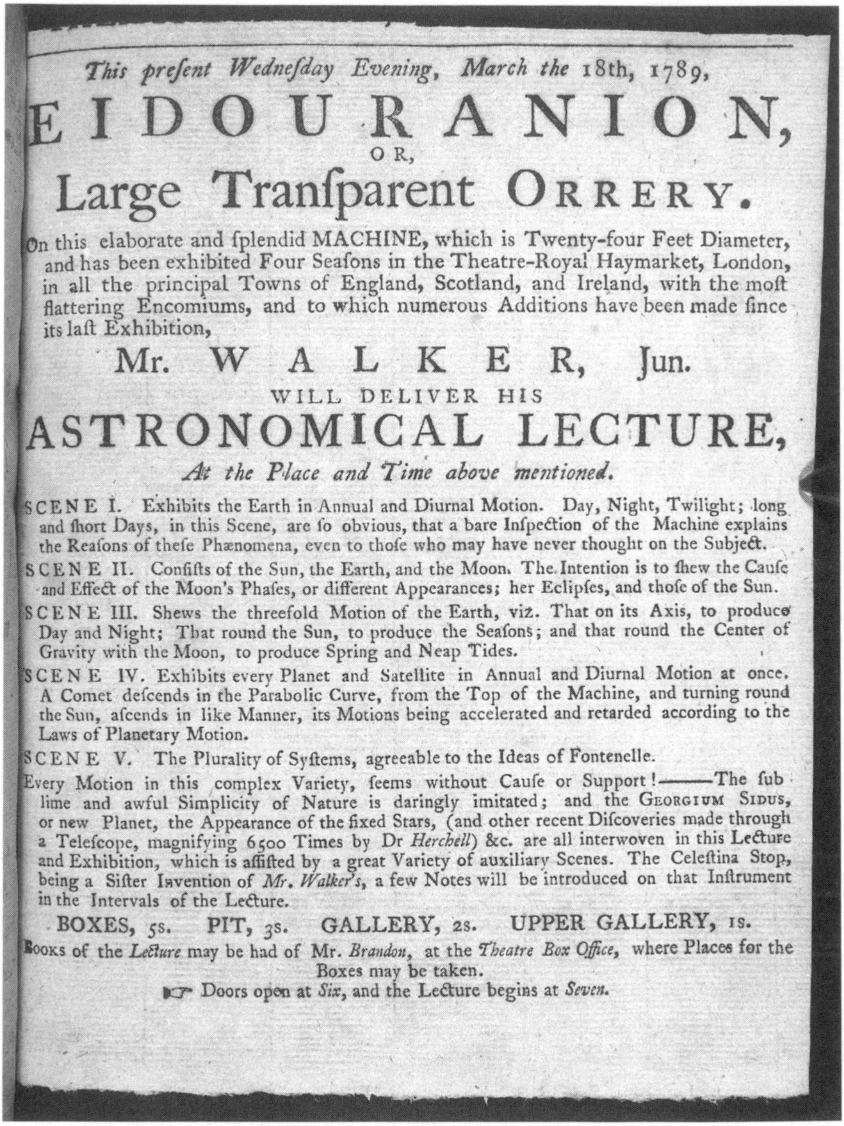

Figure 2.3. Eidouranion advertisement. (By permission of the British Library.)

tal philosophy, and on the changing status of public audiences for science in the eighteenth and early nineteenth centuries. While a substantial body of scholarship considers the role of music in the founding of experimental philosophy, focusing on figures like Robert Hooke and Isaac Newton, little attention has been paid to its role in disseminating the new science to a broader public.[3] The history of Walker's celestina thus helps us to understand how music was conceived and presented practically in relation to other fields of inquiry, and how the English makers of experimental philosophy put music to use in the advancement of knowledge during the final decades of the eighteenth century.

ADAM WALKER, "LECTURER IN EXPERIMENTAL PHILOSOPHY"

Most of what we know of Walker's early life comes from an account published in the *European Magazine, and London Review* in 1792.[4] The magazine devoted the first pages of each issue to profiles of "those persons who distinguish themselves in the service of their country, or who are made remarkable by other means"; Walker had earned his place among these ranks both by virtue of the education he delivered to the British populace and through his various inventions and publications.[5] Yet while this profile has served as a main source for subsequent writers, certain tropes and omissions make it more telling as a work of cultural mythology than of biography: it celebrates Walker (and the "lecturer in experimental philosophy" more broadly) as a pioneer committed to advancing knowledge and its practical applications for the betterment of society; it argues that he was never derivative or motivated to cater to fashion by commercial self-interest.[6] The article places special emphasis, for instance, on Walker being "self-taught"; the son of a wool manufacturer in Westmoreland, he was taken out of school early to help with the family work. By borrowing books Walker apparently managed to teach himself penmanship and accounting, skills that led to a position first at a school in Yorkshire, then one in Cheshire. At the latter he "applied himself to mathematics," in the process earning a reputation as an eccentric among the townspeople. According to the *European Magazine* chronicler, a growing sense of social isolation led Walker to abandon intellectual pursuits and "to engage in the trade of the town." The venture into trade proved a failure, demonstrating (so his earliest biographer claimed) that Walker was possessed of a philosophical disposition that rendered him unfit for such "superficial intercourse with the world." He considered turning hermit, but pressure from his friends persuaded him to take a middle path that united his intellectual predilections with social utility. Moving to Manchester, he designed a system of practical pedagogy, which his biographer praised as being "more adapted to a Town of Trade than the Monkish system still continued in our Public Schools." In 1762, he began lectures on geography and astronomy, then set up a school. For nearly five years (the account continues) he taught grammar, writing, accounting, mathematics, bookkeeping, drawing, geography, and dancing, styling himself a "teacher of the *belles lettres* in Manchester."[7] He was, in other words, cast as the very definition of a self-made man.

What the *European Magazine* failed to mention, however, was the fact that Walker's career in natural philosophy followed a model already established by earlier itinerant lecturers, such as John Arden, James Ferguson and William Griffith (or Griffis). Active since the 1740s, these men used "philosophical apparatuses"—including instruments such as orreries, air pumps, magnifying instruments, electrical machines, and steam engines—to demonstrate nature's workings in accordance with Newtonian theory. As Paul Elliott has observed, such traveling speakers often came from poor backgrounds and gained rare social mobility by studying natural philosophy; they also played a central role in popularizing science outside of London, with an emphasis on utilitarian applications particular to local concerns in industry and trade.[8] Men like Arden and Ferguson promoted the entertainment value of their lectures alongside their practical usefulness; they also emphasized their accessibility to all, even those who had never studied mathematics or indeed books of any sort. They thus espoused a novel kind of popular enlightenment: useful, universal, and made available through public lecture and demonstration. For Walker, the model had obvious relevance and appeal.

In 1766, Walker purchased a philosophical apparatus from itinerant lecturer William Griffith and began to tour northern England, southern Scotland, and Ireland (where he remained for four years), before finally settling in York. From 1766 onward, there is increased documentary evidence of Walker's activities, thanks to his newspaper advertisements and publications. Like other lecturers in experimental philosophy, he offered a twelve-lecture series on a subscription basis, beginning the course once forty or more subscribers had enrolled at one guinea each. The lectures covered a typical spectrum of philosophical topics amenable to instrumental display: mechanics, hydrostatics, pneumatics, chemistry, optics, astronomy, magnetism, electricity, and the general properties of matter.

With his 1766 publication *Analysis of a Course of Lectures on Natural and Experimental Philosophy*, Walker offered an explanation of his motives. The *Analysis* is virtually a textbook summary of enlightenment social ideology, and one that he would reiterate in many later publications. Central to his image of both his own work and the times in which he lived was his enlightened concept of "improvement"—improvement made possible, above all, by experiment and empirical observation. Walker introduced his lectures with a brief history of scientific progress: it had been scarcely 150 years (he claimed) since men had abandoned "fanciful conjectures" in favor of the only valid method for discovering "the true causes" of nature's diverse phenomena, namely "experiment, or the use of our senses." Remarkably, humanity had seen greater "advances towards perfection since the experimental method was introduced, than in the many ages before."[9] The expanding knowledge of nature and the host of practical benefits that could come from it, he argued, thus depended on what could be made evident to the senses.

And what could be made evident to the senses depended, at least in part, on instruments. Claims of possessing new and improved instruments were customary among lecturers, and by 1771 Walker could boast of several of his own invention. He was particularly proud of his transparent orrery ("more like nature than any thing of the kind") and the aforementioned musical instrument he called the celestina.[10] It is to these two inventions, and the interests and conditions that shaped them, that we now turn.

WALKER'S TRANSPARENT ORRERY: ENDLESS MOTION

By the time of Walker's "transparent" version of the instrument, the orrery had been subject to continual revision or "improvement" for over half a century. The first example is credited to John Rowley, whose model of the earth and moon orbiting the sun was made for Charles Boyle, fourth Earl of Orrery (hence the instrument's name). The device impressed early viewers with the clear idea it presented of celestial bodies in motion. On inspecting the instrument in 1713, Richard Steele enthused over its potential to enlighten all people, regardless of background. Steele made the same connections between instruments, the senses, and knowledge that later philosophical lecturers did: "It is like the receiving of a new Sense, to admit into one's Imagination all that this Invention presents to it with so much Quickness and Ease. . . . All persons, never so remotely employed from a learned Way, might come into the Interests of Knowledge, and taste the Pleasure of it by this Intelligible Method."[11] Burgeoning popular interest in Newtonian science helped create a substantial market for orreries, and instrument makers responded by designing ever more complete and complicated model solar systems.

Rowley's assistant, Thomas Wright, introduced the "grand orrery" when he added not only the remaining planets to Rowley's sun-earth-moon model but also the armillary hemisphere to illustrate Earth's major circles of latitude and longitude. This style of orrery is featured in Joseph Wright's famous painting of 1766, *A Philosopher Giving That Lecture on the Orrery, in Which a Lamp is Put in the Place of the Sun*, which shows children and adults huddled around and peering through the bronze curves of the armil-

lary hemisphere, their faces illuminated by the artificial sun. Often reproduced in histories of art and science alike, the painting has become a celebrated emblem of the rise of public science in eighteenth-century England,
its scene capturing a characteristic mix of sociability, mechanical demonstration, and religious awe.[12]

Yet as orreries became increasingly complex, they also became more difficult to understand. By the 1730s, some regarded the trend in orrery design
as one of corruption rather than improvement. John Theophilus Desaguliers,
a member of the Royal Society and popular lecturer in experimental philosophy in London, complained that instrument makers "have made improper
Additions to such Machines as have been contriv'd by Astronomers (under
pretence of Improvements) merely to make them pompous and costly; the
true Intention of the first Inventors has been destroy'd, and the Buyers have
paid dear for false Notions of Astronomy."[13] Desaguliers designed his own
orrery for use in his lectures, at once correcting the proportions between
the planets, their moons, and orbits, and eliminating superfluous additions.
What was more, where some orreries lay bare the mechanical arms and gears
that moved the planets around, Desaguliers's concealed the mechanism
within a cabinet, minimizing distractions from the clear, simple motions of
heavenly bodies (figure 2.4).

After all, as Margaret Jacob has noted, when instruments were designed
for display in a lecture setting, "simplicity and precision were essential."[14] Yet
the ideals of simplicity and precision could pull in opposite directions, re-
quiring one to be sacrificed for the other. Like Desaguliers, Walker valued the
clarity that came with concealing the orrery's mechanism. Both men, moreover, concealed their orrery's mechanism in their publications as well as
in their performances. In this they differed from lecturers-cum-instrument-
makers such as James Ferguson and Benjamin Martin, who made orreries
for sale as well as for use in their own lectures and published detailed information about the size and arrangement of wheels by which they modeled
the planets' orbits. Walker's silence surrounding the mechanics of his transparent orrery likely reflects not only the keeping of a trade secret but also a
calculated trade-off: precision in the modeling of celestial motions was sacrificed in favor of a simpler presentation in his lectures.

Because Walker left no specifications, then, the workings of his transparent orrery must remain something of a mystery. We do not know, for
example, whether the instrument was to be viewed from above like most
orreries, or whether it employed the vertical orientation that made his later
eidouranion suited to theatrical display. Nor is the meaning of the qualifier
"transparent" explained. It might refer to the technology of transparencies—

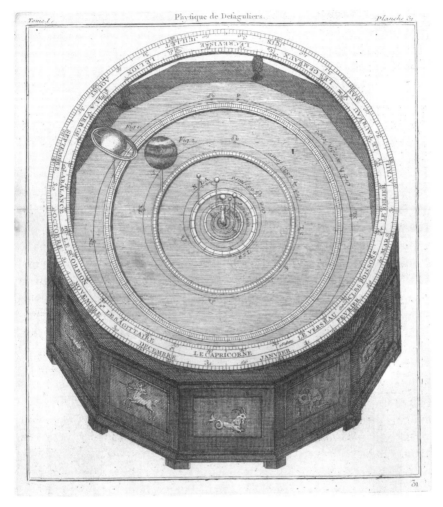

Figure 2.4. A planetarium "to shew the Motion of the heavenly bodies," from Desaguliers, *Course of Experimental Philosophy*, vol. 2 (Longman: London, 1734), n.p. (Courtesy of The Bancroft Library, University of California, Berkeley.)

painted glass illuminated from behind, a medium that was just beginning to come into vogue in the 1770s, and which in the eidouranion took the form of lighted globes that represented the planets.[15] Or it might, as Jan Golinski has suggested, signify the invisibility of the mechanism and the consequent impression that one beheld the planets supported not by wheelwork but by gravitational forces as in the heavens.[16]

What seems certain, however, is that Walker's first transparent orrery—in contrast to the eidouranion—was scaled for use in the modestly sized rooms in which he gathered his forty-person audiences: it belonged not on a stage, like the one illustrated by Burney, but on a table, like the one pictured by Gillray. As such, it took its place among the many diverse instruments Walker used for scientific demonstration in the 1770s. This collection featured not only a variety of other astronomical models—the syllabus boasts of a "variety of new constructed globes, spheres, planetariums, [and] cometariums"—but also telescopes, pneumatic devices, optical toys, and a series of model body parts: "artificial eye and ear, lungs, thorax and diaphragma."[17]

WALKER'S PATENT CELESTINA:
ENDLESS SOUND

The celestina, on the other hand, stood apart from Walker's instruments for scientific demonstration. He excluded it from the catalogue of his philosophical apparatus given in his *Syllabus of a Course of Lectures*, merely adding that "the lectures will also be enlivened by a new musical instrument, contrived by the author."[18] In other words, while the instruments listed above set "Nature . . . to work in a variety of ways, to prove the truth of her own operations," the celestina, he suggested, simply added some pleasant music to the proceedings.[19]

This has a simple explanation: music was not a subject in Walker's lectures at this time. It received only parenthetical mention within a discussion of optics, thanks to Newton's analogy between the seven colors of the rainbow and the seven pitches of the diatonic scale.[20] Sound, meanwhile, appeared briefly under the heading of pneumatics: with his "artificial ear," Walker modeled the transmission of sound from the air to the auditory nerve; and with an air pump he demonstrated that a clock striking in a vacuum could not be heard, and hence that sound cannot exist without air.[21] Matters such as the physical determinants of pitch, the mathematical basis of musical intervals, and other phenomena that might be demonstrated with musical instruments had no place in Walker's inquiries.[22]

It is thus surprising to find Walker, as a lecturer in experimental philosophy, concerned with the invention of a musical instrument. However, the "projector" Walker (to recall Burney's appellation for him) also pursued inventions for use outside the lecture room and the purpose of explicating nature. For example, he applied his knowledge of optics and lighting technologies (namely the Argand lamp) to devise revolving lighthouses for the isles of Scilly.[23] He also received a patent for "an empyreal air stove, for the purpose of purifying the air of churches, theaters, jails, sick and all other rooms, and inclosed [sic] buildings," including his own lecture rooms.[24] Like

the celestina—the only other invention for which Walker received a patent—
the empyreal air stove held dual potentials: to improve the environment of
Walker's lectures, and to find a market beyond his lecture rooms as a manufactured commodity.

The patent for the celestina was secured in 1772. At that time, Walker
described it as "a new method of producing continued tones upon an instrument." Though he would later refashion the celestina as a stop to be added
to a regular harpsichord, in the patent he described it as a new kind of keyboard instrument. According to the abridged specification, the celestina was

> a keyed instrument, shaped like a harpsichord, with one, two, or more
> wire or catgut strings to a note. The tone is produced on the strings by
> one or more threads or bands of silk, flax, wire, gut, hair, leather, &c.
> The threads or bands are kept circulating above or under the strings by a
> weight, spring, or treadle. Being pressed when in this motion against the
> strings by means of the keys, or the strings being pulled by means of the
> keys against the said threads, tones are produced from the strings as by a
> bow in the case of a violin. The tones so produced are continued as long
> as the fingers press the keys as in an organ, and are made loud or soft by
> that pressure being greater or less. This effect is also occasionally pro-
> duced by springs or weights. The celestina is also made to be played by a
> pricked barrel, as the hand or barrel organ, and is sometimes within and
> sometimes without the body of the instrument.[25]

In practice, Walker's celestina took the form of a band of silk circulated by
a treadle and pressed against the strings by the keys, this mechanism being
added to a harpsichord.

Walker described the celestina to his lecture audiences as having "all the
perfections of the organ, harpsichord, piano-forte, harmonica, or psalter,"
thereby capturing the mix of prized musical qualities to be had from a keyboard instrument equipped for sustained tone and dynamic variation.[26] In
striving to give a domestic keyboard instrument analog capacities, Walker
joined a number of other eighteenth-century inventors, but his solution was
distinct from any previously proposed. Walker's most significant predeces-
sor was Roger Plenius, a London harpsichord maker who obtained a patent
in 1741 for "a New Invention for the great Improvement of and Meliorating
the Musical Instruments called harpsichords, lyrichords . . . and spinnetts."[27]
The result was a bowed-keyboard instrument he dubbed the lyrichord; it was
advertised in newspapers as "imitat[ing] a violin, violoncello, Double-Bass,
and Organ."[28] In place of the harpsichord's string-plucking mechanism, the

lyrichord contained a set of wheels, each aligned to four strings. A string sounded when drawn down by its key to one of the wheels: machinery activated by a descending weight rotated all the wheels simultaneously but at different velocities, those for the shortest strings moving fastest to accommodate the higher frequency of vibration. Thanks additionally to a system of levered weights holding the strings in tension, Plenius was also able to claim that the instrument "never goes out of Tune." This system furthermore compensated for the variable pressure one could apply to the keys, which would otherwise change the pitch along with the volume of the sounding strings.

Walker could have learned of the lyrichord from the publications of Benjamin Martin, a fellow self-made lecturer who set up shop in London in 1740. Among the numerous works Martin printed to promote knowledge of natural philosophy (as well as his instrument-making business) was the *General Magazine of Arts and Sciences, Philosophical, Philological, Mathematical, and Mechanical* of 1755, in which the lyrichord featured among miscellaneous items intended to add relaxation and amusement to the scientific subjects (figure 2.5). The lyrichord appeared again in the second volume of Martin's *Young Gentleman and Lady's Philosophy*, first published in 1763, this time folded into a scientific discussion of "the rationale of different kinds of musical instruments," and with its device for keeping the strings in tune described as "a very curious and philosophical" contrivance.[29] These publications suggest the uncertain place of musical instruments in relation to experimental philosophy: they might be extraneous, providing mainly amusing accompaniment; they might form part of its very subject matter, demonstrating basic principles of mechanics, acoustics and music theory; or they might fall somewhere in between, as "curiosities." As a curiosity, the celestina could be a focal object of attention without being an instrument of scientific demonstration or understanding. Such was the encounter reported by a visitor to York in 1773, who "hunting after curiosities in this city . . . stumbled upon the celestina in [Mr. Walker's] lecture room, and was indeed most exceedingly surprised at the fineness of its tone, and delicacy of its expression."[30]

A comparison of Walker's celestina and Plenius's lyrichord is telling with regard to Walker's priorities. Walker reversed the action of Plenius's instrument, making it more like that of a normal harpsichord (moving the belt to the strings rather than the strings to the wheels); he reduced the mechanism from multiple wheels to a single band, and made no attempt to prevent the instrument's going out of tune. With multiple wheels rotating at different speeds, the lyrichord mechanism resembled that of a traditional orrery. In the celestina, as in the transparent orrery, Walker opted for simplification,

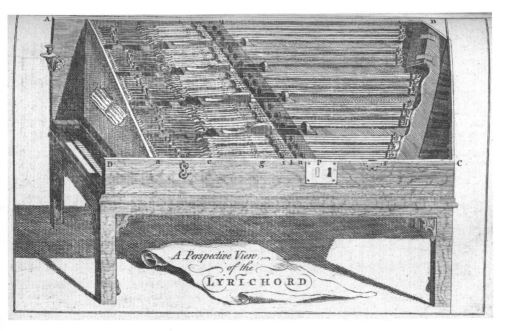

Figure 2.5. Lyrichord as first printed in Benjamin Martin, *The General Magazine of Arts and Sciences, Philosophical, Philological, Mathematical, and Mechanical* (London: W. Owen, 1755), 132–33. (By permission of the British Library.)

obtaining a striking effect without the precision of more complicated mechanisms. He sacrificed accuracy of planetary orbits in the case of the orrery; he compromised on the tone quality and intonation of the bowed keyboard.

Walker's name for his instrument further clarifies his aesthetic focus, for where the appellation "lyrichord" provided a musical/technical description of the instrument, "celestina" evoked an otherworldly music. The association of strings with celestial music was already well established: in English theater, for instance, celestial harmony was typically termed "soft music" and rendered by a string consort.[31] This theatrical tradition may help to account for the fact that Walker's celestina shared its name with another bowed keyboard instrument invented in 1761, William Mason's coelestinette (Mason's instrument was probably unknown to Walker as it was neither patented nor publicized). Rather than pedaling an endless belt, as in Walker's instrument, one played the coelestinette by drawing a bow back and forth, as on a violin. As Mason explained, the sound was produced by "the single horse-hair attached to the moveable ruler or bow, which is drawn backward and forward over the strings by the left hand of the performer, while his right is employed in pushing down the keys."[32] Walker's celestina could thus achieve something Mason's coelestinette could not: an endless sound, unarticulated by bow changes. In this endless quality, Walker's celestial sound resembled another variety of string tone that joined the celestial repertoire in the eighteenth century: that of the aeolian harp. In Benjamin Martin's "philosophical" account of musical instruments, one could also read about the aeolian harp, which was said to fill the listener "with the sensation of celestial sounds and harmony:—And it is amazing to find how the Notes will successively arise from nothing, swell to the most exalted Tones, and then gradually die away."[33] In other words, the celestial nature of the sound was not in the tone quality alone but in the behavior of the sound over time—the way it seemed to swell from nothing and die away, sustaining indefinitely, suggesting a brief auditory glimpse of the ongoing music of the spheres.

It is also tempting to find in the celestina an expression of Walker's conception of celestial mechanics. Such an interpretation would place the celestina alongside more enduring keyboard instruments of the eighteenth century. Eleanor Selfridge-Field has argued that musicians' early indifference to the fortepiano (now so-called) was due to its being "principally identified as an invention of *scientific* rather than *artistic* importance."[34] Rather than answering to a musical need, Selfridge-Field suggests, the key mechanism of the fortepiano spoke to questions of force and motion, specifically to Newton's second law, which related acceleration proportionally to applied force. Yet it is clear from Walker's positioning of the celestina vis-à-vis his philosophical apparatus that he intended its importance to be more artistic than scientific—more to do with pleasure than with knowledge. The categorical difference between the celestina and transparent orrery would be less clear, however, when the two instruments were removed from Walker's course in experimental philosophy and placed in the theater for his astronomical lectures.

EIDOTECHNOLOGY IN LONDON

After several years lecturing in the provinces, Walker ventured to London. According to the *European Magazine*'s profile, he was drawn to the city by news of Joseph Priestley's discoveries in pneumatic chemistry. When the two philosophers met, Priestley generously equipped Walker to exhibit the chemistry of gases in the Haymarket in 1778 and 1779. Walker was hesitant to take up residence in London, reportedly because he did not expect the metropolis to pay attention to "philosophical pursuits." Finally, however, he took up residence near Hanover Square, where he lectured to "numerous and genteel audiences" every winter.[35] This account of Walker's move to London sets his pure interest in knowledge and the ennobling effects of education against the commercial concerns and fashions of the metropolis. While Walker indeed maintained an important relationship with Priestley, however, he was not so exclusively concerned with the latter's chemical discoveries when he came to London, nor so reliant on his help as the *European Magazine* had its readers believe. Rather, we find Walker announcing himself in the spring of 1777 with lectures at the Artist's Exhibition Room, St. Alban's Street, Pall Mall, on electricity and optics.[36] He offered the same lectures again in 1779. With such one-off events on popular topics, Walker tested the London waters and built a reputation while maintaining his home in York.

In 1780, Walker moved his family to London and announced his full twelve-part course on experimental philosophy at a lecture room in George Street, Hanover Square.[37] The format would have been familiar to Londoners from the twelve-lecture courses taught by the likes of Benjamin Martin and James Ferguson in the 1770s.[38] Every winter thereafter until the early nineteenth century, Walker usually held his lectures Monday, Wednesday, and Friday afternoons.[39] Subscriptions remained one guinea, a sum that bought twelve transferable tickets, good for any future lecture.[40] As Walker also gave the course privately for fifty guineas, his usual audience probably remained at about forty students, as it had been in the provinces.

Yet in addition to a new audience for his lecture course, London also offered Walker new spaces and new ways of conceiving of scientific instruction. As J. N. Hays notes, "London was the nation's (perhaps the world's) center of popular exhibits, so lectures had formidable competition."[41] Walker saw this circumstance as an opportunity to diversify further, and on a grander scale. In 1782, he introduced an ambitious theater show on astronomy, which could be attended in tandem with his lecture course or on its own. For this venture, Walker built a transparent orrery larger than any before. This instrument also gained a new name: eidouranion.

The name "eidouranion" leant Walker's transparent orrery a mark of distinction, singling it out as a novel form of spectacle. It also implied kinship with a recent, highly successful addition to the London events calendar: the eidophusikon, or "image" (*eido*) of "nature" (*phusikon*). Invented by the painter and set designer Philippe-Jacques de Loutherbourg, the eidophusikon made its debut in London in February 1781. Described enthusiastically in the press as "Various Imitations of Natural phenomena, represented by Moving Pictures," and "a series of moving pictures . . . giving natural motion to accurate resemblance," the eidophusikon added the dimension of time—movement, lighting effects, and sound—to painting so as to create a more comprehensive, animated representation of natural phenomena. Where philosophical lectures deployed instruments to demonstrate nature's workings, the eidophusikon provided a model for a new class of performance: one that blended science and art in the imitation of nature. Indeed, as David Kornhaber has pointed out, scholars have struggled to classify the eidophusikon, treating it alternately as a kind of mechanical theater, a precinematic moving picture, or (Kornhaber's preference) a landmark achievement in theatrical illusionism.[42] But each of these classifications fails to account for the "philosophical" dimension emphasized by the eidophusikon's first spectators. As the *European Magazine* reported in March 1782:

> The abilities of M. de Loutherbourg, as a scene and landscape painter, were well known; it remained for him to prove, by this celebrated performance, that he was also a philosopher of the most penetrating kind, who eyed all the works of nature, and that with an observation so keen and curious, as to enable him to imitate, with the most perfect truth, her operation and phenomena.[43]

Even though de Loutherbourg's training was in painting and theater, then, his ability to reproduce natural phenomena in perfect miniature was considered a demonstration of scientific as much as artistic skill. By 1800, the eidophusikon would be joined by a number of illusionistic entertainments—including the panorama and phantasmagoria—that in deceiving the eye also offered to educate spectators about the mechanical and psychological means of trickery.[44]

The eidophusikon unfolded as a series of scenes, alternating with and sometimes accompanied by musical performance. The scenes, made up of paintings and three-dimensional objects, also featured dynamic lighting effects achieved by lamps shining onto the scene through colored glass, and transparent paintings illuminated from behind. The scenes followed the course of a day but took place all over the world, transporting spectators through space and time. Audiences for the opening season, for example, saw the following five scenes:

> 1st Aurora, or the Effects of the Dawn, with a view of London from Greenwich Park. 2d. Noon, the Port of Tangier in Africa, with the distant View of the Rock Gibraltar and Europa Point. 3d. Sun-set, a View near Naples. 4th, Moon-light, a View in the Mediterranean, the Rising of the Moon contrasted with the Effects of Fire . . . The conclusive scene A STORM and SHIPWRECK.[45]

Figure 2.6. Edward Francis Burney, *The Eidophusikon*, ca. 1782.
(© The Trustees of the British Museum.)

For the second season (1782), de Loutherbourg replaced the final storm and shipwreck with a scene of pandemonium from Milton's *Paradise Lost*; this is the scene captured in Edward Francis Burney's illustration (figure 2.6), where one enraptured audience member can be seen studying the scene with his telescope. Although Burney's illustration failed to account for the harpsichordist, musicians featured prominently in de Loutherbourg's advertisements. For the eidophusikon's first season, de Loutherbourg recruited Covent Garden composer Michael Arne to play the harpsichord, accompanying the scenes and filling the scene changes with music of his own composition: either a solo sonata or airs with his wife, the singer Ann Arne. In the second season, the Arnes were replaced by Charles Rousseau Burney (nephew of the music historian and brother of Edward Francis) and Sophia Baddeley. The form remained the same, with harpsichord music accompanying the scenes, vocal music between the scenes, and a harpsichord sonata before the grand finale.

The eidophusikon thus demonstrated not only how a major show might center around moving images of nature but also how it might incorporate music. Thomas Tolley has noted that musical performances served "to increase public interest" in the eidophusikon, but there were also significant logistical and aesthetic reasons for involving musicians in the show.[46] As the *European Magazine* reported:

> It required some time between each scene at the Eidophusicon [sic] to remove the machinery, and substitute the change; that time, though short, seemed tedious to the audience. Mr. de Loutherbourg, therefore, found it necessary to fill up that vacuity, which he did by introducing vocal and instrumental music between the scenes. Thus every moment had its amusement, and the public were satisfied.[47]

Another reviewer noted that music supplied "a relief to the sameness of the exhibition . . . so that the whole forms one of the richest, though most peculiar, feasts for the eye and eye [sic], that ever was prepared in this metropolis."[48] The match between music and scenes was also praised, as was the high quality of musicianship:

> The music attending the performance is composed by Mr. Burney, who has displayed the utmost genius, in happily suiting it to the circumstances of the different scenes. It seems the composition of Taste played by the finger of Harmony; and the beautiful little ballads sung by Mrs. Baddely [sic], with an elegant, sentimental simplicity, at the conclusion of every scene, greatly heighten the entertainment.[49]

Music was thus not merely an advertising lure, a cover for scene changes or an element of variety: it was an integral part of the performance. In the eidophusikon, "Painting and Harmony join to produce a rational and pleasing evening's entertainment."[50]

If the eidophusikon modeled the pairing of moving image and music for Walker, however, on the matter of size it was no precedent for the eidouranion. De Loutherbourg presented the show in a salon at his home on Lisle Street, Leicester Square, a space that accommodated about 130 people. The apparatus was approximately 10 feet wide, 6 feet high, and 8 feet deep. Walker's eidouranion, by contrast, was a theatrical display scaled to the largest theaters. The name "eidouranion," indeed, suggested scaling up: whereas de Loutherbourg provided an "image of nature," Walker provided an "image of the heavens"—taking spectators not just around the world but "into infinite space," not just through the course of a day but through the months and years of planetary motion.[51] The eidouranion also differed in its illusionism, presenting not an eye-deceiving copy of nature but a "new sense" (to recall Steele's early characterization of the orrery) of celestial phenomena; likewise, the celestina differed from de Loutherbourg's musical offerings in providing a sound spectators were unlikely to have encountered before.

THE EIDOURANION

While London provided key terminological and conceptual components of the eidouranion, the apparatus was born in Birmingham. In 1780, Joseph Priestley settled in that city and invited Walker to give his course there.[52] Walker presented his twelve-lecture series to Birmingham audiences in the spring of 1781, again in smaller venues; his advertisements mention the large room at the coffee house on Cherry Street and the Old Assembly Room in the square. At the same time, Walker recruited Birmingham instrument maker William Allen to execute his vision for the eidouranion. It is rarely possible to put a name to the mechanics—or "Hands," as they were called—who realized inventors' designs. In this case, however, local pride ensured that Allen's work was recognized: at the conclusion of the eidouranion's inaugural run in Birmingham, a benefit was held for Allen "for having so happily executed" the project. According to a correspondent to the *Birmingham Gazette*, the eidouranion was proof of Birmingham's superiority to other locales in matters of manufacture—it brought the "contrivances of ingenious men" together with "Hands ready and prompt to execute them."[53]

The eidouranion's Birmingham debut took place in November 1781 at the New Theatre, a two-thousand-seat venue that had opened in 1774 and would later be called the Theatre Royal. The description of what spectators would see—a series of six scenes—resembled advertisements for the eidophusikon. But rather than following the course of a day, the scenes progressed through ever more distant and complex phenomena. Walker's Birmingham advertisement also placed special emphasis on two interrelated aspects of the display, its instructional value and the invisibility of its machinery:

> 1st. The Ptolomaic System of the Universe in Transparency.—2dly. That Part of the Copernican System which relates to the Sun's Motion on his Axis. The Earth's Motions, both annual and diurnal; shewing thereby, how Day and Night, long and short Days, the Seasons, &c., are produced; so evident and like Nature, that a bare inspection of the Machine, and a Quarter of an Hour's attention, may give the most Ignorant a clear Idea

of these Phenomena.—3dly. The Motion and Phases of the Moon; with Eclipses of both Sun and Moon.—4thly. The Earth and Moon's Motion round their common Centre of Gravity; and how Spring and Neap Tides are produced.—5thly. A Transit of Venus.—6thly. A Grand Display of the whole Solar System, viz., the Sun, Mercury, Venus, the Earth and Moon, Mars, Jupiter, Saturn, and their Satellites, all in motion without any visible Machinery; Together with the Parabolic Descent and Ascent of a Comet.[54]

The elimination of mechanism in the service of clarity and the creation of spectacle in the service of instruction would remain central themes in Walker's promotion of the eidouranion in the coming years.

When Walker returned to London that winter, he offered his twelve-part lecture series at his George Street home and his new Astronomical Lecture on the eidouranion at Covent Garden and in the Haymarket. London advertisements described the eidouranion as a machine twenty feet in diameter, with an earth so large that the effects "upon the smallest Island may be distinctly seen in the most distant parts of the Theatre."[55] They promised to render "astronomical truths so plain and intelligible, that even those who have not so much as thought upon the subject, may acquire clear ideas of the laws, motions, appearances, eclipses, transits, influences, &c. of the planetary system."[56] But viewers were also invited to marvel at Walker's ingenuity in creating such an "elaborate and splendid MACHINE," even as the "stupendous Orrery exhibits without any apparent machinery or support."[57] In a move that would seem counter to the enlightening mission of experimental philosophy but that served at once to protect the illusion of viewing the heavens and frustrate would-be imitators, there was "no admittance behind the scenes."[58] The eidouranion was so successful in London that its exhibition continued for an unprecedented fifty years, with Adam Walker's sons William and Dean eventually replacing him in the role of lecturer.

Initially, the celestina seems not to have played a role in the astronomical lecture on the eidouranion: early advertisements made no mention of it, nor of any other music.[59] In the eidouranion's second year (1783), however, an advertisement announced that "the CELESTINA lately introduced, has met the Approbation of the Public," and thus "it will as usual be continued in the Intervals of the Lecture."[60] By 1789 the Walkers had settled upon language that would be used to the end of the eidouranion's half-century run. This language emphasized the kinship of the musical instrument with the astronomical: "the celestina stop, being a sister invention of *Mr. Walker's*, a few Notes will be introduced on that Instrument in the Intervals of the Lecture."[61] Unlike de Loutherbourg, the Walkers never named the performing musicians, instead keeping the focus on the invented instrument. The conceit of the "sister invention" (to which there are no allusions at the time Walker invented the celestina) suggests a new relationship between music and science: a rediscovery of the sister relationship between music and astronomy dear to the ancients but excluded from Walker's modern system of natural philosophy, updated to have its basis not in mathematical theory but in the sensible results of instrumental practice.

But there were other, more practical reasons for equating the celestina and eidouranion: the Walkers wanted to promote the instrument's sale for domestic use. In 1783, they began to advertise their "Celestina Stop" to the London public, now construing the device as an addition to the harpsichord rather than a new instrument in the shape of a harpsichord. Emphasizing musical expression over celestial thoughts, they explained:

> The effects this improvement produce on the Harpsichord, are, a continuation of tone, swell and dimminuendo, [sic] with the Piano and Forte by the pressure of the finger; hence the grand effects of the Organ, with the delicacy of the Musical Glasses, or Viol d'Amor, are given to the Harpsichord, and a degree of musical expression superior to most instruments. It has the most enchanting effect as an accompaniment to the voice in pathetic or sentimental singing. It can be introduced as a solo instrument in the middle movements of Harpsichord Concertos, giving a pleasing relief to the ear with the most striking contrast.

The Walkers suggested that "the lovers of music" could hear the celestina for themselves at No. 8, Great Pulteney Street, Golden Square, "where an instrument is always ready for their inspection."[62]

This location in the heart of Soho put Walker's celestina in close proximity to many of London's harpsichord and piano makers, including Joseph Mahoon, William Stodart, Kirkman, and the firm of Shudi and Broadwood. Indeed, the Walkers seem to have developed a business relationship with the last of these: the celestina is mentioned in Shudi and Broadwood workbooks as early as 1775, and the firm made payments to "Walker & Co."[63]

Thanks to one famous customer, we have some record of the celestina's life beyond the Walkers' lectures. Thomas Jefferson visited London in 1786, staying near the city's keyboard makers at 14 Golden Square. Shortly after his return to America, he sought a harpsichord from Kirkman for his daughter

Martha, to be made with a Venetian swell and celestina stop. He asked that the celestina be operated by clockwork rather than treadle, so as to eliminate the constant foot-pedaling that he found "diverts the attention and dissipates the delirium both of the player and hearer." Kirkman, however, objected to the incorporation of the celestina stop, claiming that rosin on the silk band "not only clogs the wheels and occasions it to be frequently out of order, but in a short time, adheres so much to the strings as to destroy the tone of the instrument."[64] At that time, Charles Burney—Jefferson's intermediary with Kirkman—was not sufficiently acquainted with the celestina stop to determine how much Kirkman's objections were based in truth, and how much they owed to the prejudices of a business rival. Jefferson, however, reasoned that the problem of rosin build-up on the strings could be dealt with by "wiping the strings from time to time," and by using the stop sparingly. He would not forego the stop, for "in the movements to which it is adapted I think its effect too great not to overweigh every objection"; once Kirkman finished all other aspects, the harpsichord was sent to the Walkers for the addition of the celestina stop.[65] Having tested the completed instrument, Burney noted that Walker had improved the stop: the bow was easier to bring into contact with the strings, making the device suitable for more than "mere Psalmody, as was the Case at the first invention"; and the bow was less likely to produce a "*Scream*" when the keys were pressed too firmly. Heard from a short distance, the celestina stop reminded Burney of "the best and most expressive part of an organ, the Swell."[66] But Kirkman might have been right after all about the celestina stop's deleterious effects on its host instrument: by 1825 the Jefferson harpsichord was deemed unplayable and destroyed.[67]

The 1783 advertisement and Jefferson's correspondence suggest what may have been played on the celestina. Slow movements were especially suited to it; indeed, the instrument proved unable to cope with faster passages. The Walkers specified the "pathetic" and "sentimental" as expressive domains in which the celestina excelled, while Jefferson noted its ability to induce "delirium" in the performer (*sans* foot pedal). It is noteworthy how at odds these aesthetic registers are with that of the eidouranion—a grand spectacle of a sublime science devised to enlighten as well as entertain. The harpsichord with celestina stop was an intimate device that, unlike the transparent orrery, was not—and indeed, probably could not be—recalibrated for a theatrical scale. In other words, the eidouranion and celestina were paired better in advertising copy than they were in performance.

"PROGRESS OF THE ARTS":
DISENCHANTING THE SPHERES

The astronomical lectures on the eidouranion remained a staple of both London and provincial performance calendars through the 1820s, delivered mainly by Walker's sons while Adam focused on his lecture courses (and continuing some years after the father's death in 1821).[68] Until about 1800, the Walkers gave the only astronomy lecture on offer in theaters. By the 1820s, however, numerous rivals had emerged, among them Robert Evans Lloyd, who coined the name "dioastrodoxon" for a very similar "transparent orrery"; and a Mr. Bartley, who illustrated his astronomical lectures with magic lantern slides and called his show the Uranologia. The Walkers' theatrical performances were also adapted for domestic consumption, making a return to the kind of setting where Adam Walker got his start: the optical instrument maker Charles Blunt advertised a set of magic lantern slides and accompanying volume as "a complete eidouranion on a small scale, for the amusement and instruction of a family circle, or the higher classes of schools."[69] These imitators and the continued use of music attest to the success of the Walkers' formula. Lloyd went so far as to employ the very same musical instrument, advertising that his dioastrodoxon was "accompanied by the dulcet notes of the Celestina."[70]

Once a triumph of industriousness and innovation, the eidouranion thus became a matter of maintenance—of carrying on an established form in competition with newcomers. In the nineteenth century, the eidouranion "season" was reduced to Lent, when it benefited from the ban on dramatic performances in theaters, and therefore joined other such "classics" as Handel's *Messiah*. As a Lenten offering, the eidouranion lectures interspersed explanations of celestial mechanics with large doses of poetry celebrating the divine maker; they also began to incorporate newly canonical oratorio music. According to an 1819 advertisement, "an Air or Hymn of Handel or Haydn" was performed after each scene. These included Handel's "Holy, Holy, Lord," from the *Messiah*, and Haydn's duet "The Heavens are telling" from *The Creation*.[71]

Though they turned to the past for musical repertoire, Walker and his sons kept their lectures up to date with the latest scientific discoveries. By 1789, for example, the eidouranion included "the Georgium Sidus, or new Planet, the distance of the fixed Stars (and other recent Discoveries made through a Telescope, magnified 6500 Times, by Dr. Herchell [sic])."[72] Adam Walker advertised his lecture courses as including "every material modern discovery and improvement to the present time," and his accompanying publications became lengthier with the addition of new topics and theories. This principle also extended to musical instruments: the eidouranion's 1819 season briefly featured the newly invented oedephone, an imitation wind band performed by its touring inventor, the Viennese instrument maker Vandenburgh. The oedephone's cameo appearance suggests that the eidouranion provided a space for the demonstration of experimental musical instruments. To the end, the Walkers continued to use the celestina, that modified harpsichord born of an eighteenth-century acoustic image of celestial harmony scaled to the intimate salon. Although elsewhere it was possible to promote the celestina as a "new musical instrument," for London audiences of the 1820s it performed the sound of a bygone era.[73]

Yet the aging celestina still offered a host of new expressive possibilities. When demonstrating the advancement of knowledge, lecturers often began by recollecting old, discredited ideas. In the early years of the eidouranion, the Walkers first presented Ptolemy's geocentric model of the universe, "with the Planets and fixed Stars in motion, agreeable to that *erroneous* Hypothesis." The false system of the ancients was corrected in the second scene, which "Exhibits the Earth and Sun, according to the Copernican or *true* system."[74] By the 1790s, the Walkers had dropped the Ptolemaic system from their scenes in order to make room for new discoveries, such as Herschel's discovery of Uranus. But rather than disappearing altogether, discredited hypotheses shifted from the visual to the auditory register of the lectures— from the image to the sound of the heavens. Evidence for use of the celestina to illustrate "erroneous" astronomy comes mainly from literary accounts of the eidouranion: as the show aged, it attracted interest, not as much in the popular press but in works of fiction. In Maria Edgeworth's *Frank: A Sequal* [sic] *to Frank in Early Lessons* (1822), for instance, a family attends a lecture on the eidouranion. As the curtain rises to reveal "Globes that seemed self-suspended in air," the audience hears soft music "from an harmonica, which was concealed behind the scenes."[75] The lecturer then emerges, and begins "to talk of *celestial harmony,* or *the music of the spheres,* which he told them they had just heard: yet which had never really existed, except in the fanciful systems of the ancients."[76]

Despite several updates, by the mid-1820s the eidouranion had acquired "an air of times gone by": such was the observation of the volume *London Lions for Country Cousins and Friends about Town* (1826), which commemorated the "growing splendour and increasing magnificence" of the "British Metropolis." According to the *London Lions,* astronomical lectures held a preeminent place among London's "improvements and amusements," com-

bining as they did the most effective device for the dissemination of knowledge (the public lecture) with the most sublime science (astronomy), and
thereby expressing the character of an age "proudly distinguished by a general
cultivation of science, and a love of literature and useful knowledge, which
spreads over the whole intelligent community." While recognizing Adam
Walker's role in fostering such an age, however, the author found that the
Walkers' lectures paled in comparison to the "improved lecture of Bartley."
Strikingly, the *London Lions* faulted the Walkers for failing to keep up not
with astronomical science, but with music and the scenic arts: "half a century of general advancement in every art which ornaments society, finds the
exhibition of the eidouranion so little assisted by auxiliaries, which would,
unquestionably, render it doubly delightful, a thousand times more useful."[77]

The eidouranion could be dismissed merely on account of its outdated
"auxiliaries," though, because the very notion of "improvement" on which
it relied belonged to a previous age. As we have seen, Walker believed that
humankind was advancing steadily toward perfect knowledge. This applied
to all people, regardless of station or background: thanks to the nature of
sensory evidence, everyone could recognize the truth of the modern natural
philosophy, and learn its basic principles and facts. What was more, public
access to knowledge was an essential part of the scientific process, crucial
to "rendering those sciences more perfect." As Simon Schaffer has observed,
Walker's was an "epistemology which favored the community of all witnesses of nature" as the guarantor of truth.[78]

Walker expressed such ideas in a 1792 letter on the "progress of the arts"
at Oxford University, wherein he compared the states of music and astronomy at that fine institution. Music, he found, had made great "strides towards
perfection"; where once vulgar songs were the standard, now Handel was
revered and his music performed by thousands at the university. The "improved taste in that divine science" was above all demonstrated by the degree
recently granted to Haydn, "this musical Shakespeare, this musical Drawcan-
sir, who can equal the strains of a Cherub, and enchant in all the gradations
between those and a ballad—a genius whose versatility comprehends all the
powers of harmony!" In astronomy, however, Walker found no such progress: "Would to God I could say as much for the science I love!" He asked:
"Must Astronomy, which recognizes the whole universe, be more limited
in the liberality of its Professors than the sensual Arts?" The trouble was an
old, Monkish seclusion—a denial of access to knowledge. The advances of
Oxford scientists were not shared with the public, and Walker himself had
been turned away from the university observatory.[79]

In the early nineteenth century, the gap between the popular and professional sciences that Walker lamented only grew wider, and by 1850, as Ian Inkster has observed, "the link between intellectual advance and the wider astronomical culture was no longer visible."[80] This shift is observable in audiences for astronomy, in the striking contrast between the group reverently gathered around the orrery in Wright's painting of 1761, or around the cidophusikon in 1782, and the theatrical audience in Burney's eidouranion illustration of 1817. The latter seems less a "community of all witnesses of nature" than an unruly mass ready to be dazzled by sights and sounds (and equally ready to be distracted by their neighbors). With public performance severed from scientific progress, it made sense to measure the metropolis not by the musical tastes of its populace, or by their access to scientific apparatus, but merely as the *London Lions* did, by the splendor of their entertainments. The fate of these musico-astronomical exhibitions, in other words, bears witness not only to the well-known story of the increasing "illiberalism" of scientific practice but also to a second disenchantment of astronomical knowledge. Where Walker had discarded harmonic models of the cosmos yet joined music and astronomy through his inventions, now new specialisms perpetuated a new order of things, and a universe beyond the pleasurable revolutions of theatrical seasons and mechanical instruments.

NOTES

1. "Philosophical Arcana," *Morning Chronicle and London Advertiser*, January 5, 1782.

2. "To Thomas Jefferson from Charles Burney, 20 January 1787," *Founders Online*, National Archives, http://founders.archives.gov/documents/Jefferson/01-11-02-0061.

3. On the role of music in the founding of experimental philosophy, see especially Penelope Gouk, *Music, Science, and Natural Magic in Seventeenth-Century England* (New Haven, CT: Yale University Press, 1999).

4. "Mr. A. Walker, Lecturer in Experimental Philosophy," *European Magazine, and London Review*, June 1792, 411–13.

5. "Introduction," *European Magazine, and London Review*, January 1782, iii. On the *European Magazine*, see Alvin Sullivan, ed., *British Literary Magazines*, vol. 1 (Westport, CT: Greenwood Press, 1983), 106–9.

6. Later accounts employing information from the *European Magazine* profile include Simon Schaffer, "Natural Philosophy and Public Spectacle in the Eighteenth Century," *History of Science* 21 (1983): 1–43; and Jan Golinski, *Science as Public Culture: Chemistry and Enlightenment in Britain, 1760–1820* (Cambridge: Cambridge University Press, 1992).

7. Adam Walker, *Analysis of a Course of Lectures on Natural and Experimental Philosophy* (Kendal: Printed for, and sold by the author, 1766), t.p.

8. Paul Elliott, "The Birth of Public Science in the English Provinces: Natural Philosophy in Derby, c. 1690–1760," *Annals of Science* 57 (2000): 61–100.

9. Walker, *Analysis of a Course of Lectures*, 3.

10. Adam Walker, *Syllabus of a Course of Lectures on Natural and Experimental Philosophy* (Liverpool: 1771), 2.

11. Quoted in Henry King, *Geared to the Stars: The Evolution of Planetariums, Orreries, and Astronomical Clocks* (Toronto: University of Toronto Press, 1978), 154.

12. See Elliott, "The Birth of Public Science"; Elizabeth E. Barker, "New Light on the 'Orrery': Joseph Wright and the Representation of Astronomy in 18th-Century Britain," *British Art Journal* 1 (2000): 29–37.

13. Desaguliers, *A Course of Experimental Philosophy* (1734); quoted in King, *Geared to the Stars*, 171.

14. Margaret Jacob, *Practical Matter: Newton's Science in the Service of Industry and Empire, 1687–1851* (Cambridge, MA: Harvard University Press, 2004), 89.

15. See John Plunkett, "Light Work: Feminine Leisure and the Making of Transparencies," in *Crafting the Woman Professional in the Long Nineteenth Century: Artistry and Industry in Britain*, ed. Kyriaki Hadjiafxendi and Patricia Zakreski (Farnham, UK: Ashgate, 2013), 43–68.

16. Jan Golinski, "Sublime Astronomy and the End of the Enlightenment: Adam Walker and the Eidouranion," paper delivered at the Center for Science, Technology, Medicine & Society colloquium series, University of California, Berkeley, February 9, 2012.

17. Walker, *Syllabus of a Course of Lectures*, 2.

18. Ibid., 3.

19. Adam Walker, *A System of Familiar Philosophy in Twelve Lectures*, vol. 1 (London: G. Kearsley, 1802), vi.

20. Walker, *Analysis of a Course of Lectures*, 64.

21. Ibid., 28.

22. In the 1790s, Walker expanded his discussion of sound under pneumatics to include phenomena such as musical vibrations, the aeolian harp, and sound conduction.

23. Walker, *System of Familiar Philosophy*, 2:141.

24. *Patents for Inventions: Abridgments of Specifications Relating to Ventilation, A.D. 1632–1866* (London: George E. Eyre and William Spottiswoode, 1872), 4. Walker's stove was awarded a patent in 1786.

25. *Patents for Inventions: Abridgments of Specifications Relating to Music and Musical Instruments, A.D. 1694–1866* (London: George E. Eyre and William Spottiswoode, 1871), 8.

26. Walker, *Syllabus of a Course of Lectures*, 3.

27. Bennet Woodcroft, *Reference Index of Patents of Invention, from 1617 to 1852* (London: George Edward Eyre and William Spottiswoode, 1855), 82. See also Eric Halfpenny, "The Lyrichord," *Galpin Society Journal* 3 (1950): 46; Donald Boalch, *Makers of the Harpsichord and Clavichord, 1440 to 1840* (New York: Macmillan, 1956), 85–86. As Halfpenny notes, it is misleading to say—as Boalch and others have—that Plenius patented the lyrichord; rather, he patented a set of devices to improve existing keyboard instruments, some of which he subsequently employed in the new instrument he called a lyrichord.

28. "Roger Plenius, The Inventor of the Lyrichord" [classified ad], *London Evening Post*, March 3–5, 1747.

29. Benjamin Martin, *The Young Gentleman and Lady's Philosophy*, vol. 2 (London: W. Owen, 1763), 382.

30. Philo Musicus, "To the Editor," *The York Chronicle; And Weekly Advertiser*, March 19, 1773, 108.

31. John Manifold, "Theatre Music in the Sixteenth and Seventeenth Centuries," *Music & Letters* 29, no. 4 (1948): 366–97.

32. J. Mitford, ed., *The Correspondence of Horace Walpole, Earl of Orford, and the Rev. William Mason*, vol. 1 (London: Richard Bentley, 1851), 432.

33. Martin, *The Young Gentleman and Lady's Philosophy*, 2:385.

34. Eleanor Selfridge-Field, "The Invention of the Fortepiano as Intellectual History," *Early Music* 33 (2005): 83.

35. "Mr. A. Walker, Lecturer in Experimental Philosophy," 412.

36. "Mr Walker's Lecture" [classified ad], *Morning Post and Daily Advertiser*, May 22, 1777.

37. "Mr. Walker's Lecture on Experimental Philosophy . . ." [classified ad], *Morning Chronicle and London Advertiser*, February 15, 1780; he also advertised in the *Gazetteer and New Daily Advertiser*, January 4, 1780.

38. John R. Millburn, "The London Evening Courses of Benjamin Martin and James Ferguson, Eighteenth-Century Lecturers on Experimental Philosophy," *Annals of Science* 40 (1983): 437–55.

39. Advertised in William Walker, *An Epitome of Astronomy* (Bury: printed for the author by P. Gedge, 1798, 1800).

40. "Philosophical Arcana," *Morning Chronicle and London Advertiser*, January 5, 1782.

41. J. N. Hays, "The London Lecturing Empire, 1800–50," in *Metropolis and Province: Science in British Culture, 1780–1850*, ed. Ian Inkster and Jack Morrell (Philadelphia: University of Pennsylvania Press, 1983), 110.

42. David Kornhaber, "Regarding the Eidophusikon: Spectacle, Scenography, and Culture in Eighteenth-Century England," *Theatre Arts Journal* 1 (2009): 45. See also Charles Baugh, "Philippe de Loutherbourg: Technology-Driven Entertainment and Spectacle in the Late Eighteenth Century," *Huntington Library Quarterly* 70 (2007): 251–68; Simon During, "Beckford in Hell: An Episode in the History of Secular Enchantment," *Huntington Library Quarterly* 70 (2007): 269; and Richard D. Altick, *The Shows of London* (Cambridge, MA: Belknap Press, 1978), 119–24.

43. "A View of the Eidophusikon," *European Magazine*, March 1, 1782, 180. The panorama similarly merged art and science in precision imitation; see John Tresch, *The Romantic Machine: Utopian Science and Philosophy after Napoleon* (Chicago: University of Chicago Press, 2012), 136–37.

44. See Iwan Morus, "Seeing and Believing Science," *Isis* 97 (2006): 101–10.

45. Advertisement in the *Morning Herald*, March 14, 1781, cited in Ralph G. Allen, "The Eidophusikon," *Theatre Design and Technology* 7 (1966): 12.

46. Thomas Tolley, *Painting the Cannon's Roar: Music, the Visual Arts and the Rise of an Attentive Public in the Age of Haydn* (Burlington, VT: Ashgate, 2001), 297.

47. "Anecdotes of Mr. de Loutherbourg," *European Magazine*, March 1, 1782, 182.

48. "Guildhall Intelligence," *Morning Chronicle and London Advertiser*, February 28, 1781.

49. "As the Much-Admired Eidophusikon of the Celebrated Mr. De Loutherbourg. . ." *London Courant, Westminster Chronicle and Daily Advertiser*, February 6, 1782.

50. Ibid.

51. William Walker, *An Account of the Eidouranion*, 10th ed. (London, 1793), 23.

52. Priestley to Walker, 1780, quoted in Golinski, *Science as Public Culture*, 99.

53. John Alfred Langford, *A Century of Birmingham Life*, vol. 1 (Birmingham, UK: E. C. Osborne, 1868), 252–53.

54. "Mr. Walker's Astronomical Lecture on the Eidouranion," *Birmingham Gazette*, November 5, 1781, quoted in ibid., 253.

55. "Mr. Walker's Astronomical Lecture on the Eidouranion" [classified ad], *London Courant, Noon Gazette, and Daily Advertiser*, January 28, 1782.

56. Walker, *An Account of the Eidouranion*, 4.

57. Eidouranion broadside, March 18, 1789. Theatre Royal, Covent Garden, Westminster, "This present Wednesday evening, March the 18th, 1789, Eidouranion, or, large transparent orrery. . . . Mr. Walker, jun. will deliver his astronomical lecture, at the place and time above mentioned" ([London], [1789]), *Eighteenth Century Collections Online*, Gale, UC Berkeley, http://find.galegroup.com/ecco/infomark.do?&source=gale&prodId=ECCO &userGroupName=ucberkeley&tabID=T001&docId=CW3313600683&type=multipage& contentSet=ECCOArticles&version=1.0&docLevel=FASCIMILE; "Mr. Walker's Astronomical Lecture on the eidouranion" [classified ad], *London Courant*, January 28, 1782.

58. "Mr. Walker's Astronomical Lecture on the Eidouranion" [classified ad], *London Courant*, January 28, 1782.

59. Walker also omitted the celestina from advertisements for his courses, so it is unclear whether he continued to use the instrument.

60. "Eidouranion, The Candour and Approbation hitherto shewn Mr. Walker" [classified ad], *Felix Farley's Bristol Journal*, February 22, 1783.

61. Theatre Royal, Covent Garden, Westminster, "This present Wednesday evening."

62. "Celestina Harpsichord" [classified ad], *Morning Herald and Daily Advertiser*, March 8, 1783.

63. David Wainwright, *Broadwood by Appointment: A History* (London: Quiller, 1982), 53–55. Thanks to Margaret Debenham for sharing source materials on the Walkers' dealings with harpsichord makers.

64. "To Thomas Jefferson from John Paradise, with Enclosure, 27 June 1786," *Founders Online*, National Archives, http://founders.archives.gov/documents/Jefferson/01-10-02 -0014.

65. "From Thomas Jefferson to Charles Burney, 10 July 1786," *Founders Online*, National Archives, http://founders.archives.gov/documents/Jefferson/01-10-02-0048.

66. "To Thomas Jefferson from Charles Burney, 20 January 1787," *Founders Online*, National Archives, http://founders.archives.gov/documents/Jefferson/01-11-02-0061 (original emphasis).

67. A Kirkman harpsichord in the Andreas Beurmann collection of historical keyboard instruments is the only known surviving instrument with a celestina stop. The instrument was built in 1768; how it came to be outfitted with the celestina is undocumented. Andreas Beurmann, *Harpsichords and More: Harpsichords, Spinets, Clavichords, Virginals* (New York: Georg Olms, 2012), 224–25.

68. King, *Geared to the Stars*, 311.

69. "An Improved Magic Lantern" [advertisement], *Repository of Arts, Literature, Commerce, Manufactures, Fashions and Politics* 12 (1 January 1814), 4.

70. "Astronomy, Accompanied by the dulcet notes of the Celestina" [classified ad], *The Edinburgh Literary Journal; or, Weekly Register of Criticism and Belles Lettres*, January 1829, 20.

71. Broadsides for "Mr. D. F. Walker's Lecture, on his entirely New Eidouranion," in

[*A collection of play bills of the New Theatre royal, English Opera House, Strand. 1819–33*], held by University of Oxford Bodleian Library, shelfmark Vet. A6 c. 163–64.

72. See the broadsheet reproduced in figure 2.3. On William Herschel's astronomical discoveries (including of the "new Planet" Uranus mentioned here), see Richard Holmes, *The Age of Wonder: How the Romantic Generation Discovered the Beauty and Terror Science* (New York: Pantheon Books, 2008). Holmes also discusses Herschel's musical background and the practiced use of instruments and skilled "reading" (scores or starry skies) that transferred from music to astronomy.

73. "Account of Mr. Walker's New Musical Instrument Called the Celestina," *Glasgow Mechanics Magazine*, May 22, 1824, 329.

74. "Mr. Walker's Astronomical Lecture on the Eidouranion" [classified ad], *London Courant*, January 28, 1782 (original emphasis).

75. Maria Edgeworth, *Frank: A Sequal to Frank in Early Lessons*, vol. 2. (New York: William B. Gilley, 1822), 52.

76. Other nineteenth-century accounts confirm that the Walkers used the celestina to give "an idea of the music of the spheres," and that it sounded when the curtain rose; see "The Eidouranion," *Monthly Mirror*, March 1808, 275; "On the Astronomical Machine, Called the Eidouranion; or Large Transparent Orrery," *Glasgow Mechanics Magazine*, January 10, 1824, 20; and "Account of Mr. Walker's New Musical Instrument Called the Celestina," *Glasgow Mechanics Magazine*, May 22, 1824, 329.

77. Horace Wellbeloved, *London Lions for Country Cousins and Friends About Town* (London: William Charlton Wright, 1826), 2.

78. Schaffer, "Natural Philosophy and Public Spectacle in the Eighteenth Century," 26.

79. Adam Walker, *Remarks Made in a Tour from London to the Lakes of Westmoreland and Cumberland* (London: 1792), 7–10.

80. Ian Inskter, "Advocates and Audience—Aspects of Popular Astronomy in England, 1750–1850," *Journal of the British Astronomical Association* 92 (1982): 123.

CHAPTER THREE

Transparent Music and
Sound-Light Analogy ca. 1800

ELLEN LOCKHART

SCRUTON-IZED OR *CORDER*-ED?

Musicology has a program-music problem—a paradox, really. On one hand, we are invested in musical signification, in discovering what and how music means. On the other, we find overt or articulated musical references to outside objects, such as word painting, embarrassing; and we often assign a lower status to genres that determine their own interpretation. Take, for instance, the article "Programme Music" in the *New Grove Dictionary of Music and Musicians*, which is among its most polemical offerings. The editorial board solicited "programme music" and two other related entries— "Absolute Music" and "Expression"—not from a career musicologist but from the English philosopher Roger Scruton. For the *New Grove* Scruton defined program music as "of a narrative or descriptive kind"—in other words, music that either tells a story or paints a scene. What is more, Scruton emphasized that music with a program must forsake music's "autonomous principles" to behave in ways that are appropriate to that aim.[1]

Scruton left little doubt that he found program music interesting primarily—perhaps even only—because he could define autonomous or "absolute" music against it. But within the category of program music there were further hierarchies of value to be established, particularly that he preferred narrative music to music that describes. This was predetermined by his method: Scruton aimed to define program music ontologically while, at the same time, giving priority to claims that were advanced when the category was created. Franz Liszt had coined the term, and put its principles to use in several "symphonic poems" with literary sources: *Tasso* (1849), *Hamlet* (1858), and *Two Episodes from Lenau's Faust* (ca. 1860). Scruton suggested that Liszt himself did not regard music as a means of representation, *per se*, despite substantial evidence to the contrary. Rather, the purpose of program music was to "put the listener in the same frame of mind as could the objects themselves" (this is Scruton ventriloquizing Liszt). In its literary subgenre, program music did not mimic external phenomena directly, but engaged with them at the safe distance ensured by a mediating textual layer. Scruton traced this idea back to Jean-Jacques Rousseau, in particular his notion of a shared ancestor of music and language.

Rousseau's and many other eighteenth-century writings in this vein can hardly assume parentage of program music: if they are asked to do so, the category threatens to merge with any other theory of musical expression. But the purpose of Scruton's lineage was to rescue a few, select pieces of program music—by Beethoven, Berlioz, Mendelssohn, Schumann, Richard Strauss, and Debussy—from the more disreputable form of musical representation that he called "naïve pictorialism," or tone painting. Such was the haste to do so that he tripped over his own chronology, quoting by the end of the first paragraph Beethoven's affirmation that his Pastoral Symphony was "mehr Ausdruck der Empfindung als Malerey" (more the expression of feeling than painting). In this sense, Scruton's language echoes that of Theodor Adorno, who decried the "crass infantilism" of tone-painting and warned against composers who "treat time as if it were a cartoon." Both writers' rhetoric bears traces of the modernist abhorrence of Walt Disney's *Fantasia*, that middlebrow embarrassment seeking to educate "ordinary folks" in "so-called classical music" (Disney's own words) by providing canonic works with memorable visual referents.[2]

An approach very different from Scruton's—and a salutary lesson in the mutability of musicological priorities—can be found in the article "Programme-Music" in Sir George Grove's first *Dictionary of Music and Musicians*. The author was Frederick Corder, an English composer and music historian.[3] Corder averred that "employ[ing] music to imitate the *sounds* of nature" is a "degradation" of the art. He objected to musical figures imitating birdsong, trumpet calls, thunderclaps, and so on. However—and here is the crucial distinction—music should nonetheless use "every means within its power" to direct the listener's mind toward the appropriate ideas. What is more, the locus of these ideas was an imaginary organ, the "mind's eye"—a notion entirely absent from Scruton's account. For Corder, the more refined the listener, the greater the number of "vivid pictures" that appear in his mind as he listens. Such pictorialisms are legitimate because they "assist the mind which is endeavoring to conjure up the *required* images." Since tonepainting helps "the uninitiated" to make sense of "a chaos of sound," program music becomes "the noblest" branch of its art.

Corder's canon was thus rather different from Scruton's. Corder dismissed Berlioz's and Liszt's "leitmotivic symphonies" as failures of the programmatic genre. More successful were Sterndale Bennett's *Paradise and the Peri* Overture, or Hans von Bülow's ballad for orchestra op. 16, *Des Sängers Fluch*. For Corder, program music had its origins in fifteenth-century word painting; the ancestors of *Mazeppa* included the battle sounds and bird songs in the "Dixième livre des chansons" (Antwerp, 1545), the "miaous" in Adam Krieger's vocal fugue of 1667, and the "Fantasia on the Weather" in the Fitzwilliam Virginal Book. (Scruton would have regarded all three as irrelevant: the first two because they have sung text, the last because it lacks a narrative element.) In short, the account of program music in Grove's first dictionary sets Beethoven, Liszt, and Richard Strauss cheek-by-jowl with a host of other composers whose ability to stimulate the "mind's eye" was, in Corder's view, equal or even superior.

MUSIC ON AN ENTIRE NEW PLAN

One such composer was Nicola Sampieri, whom Corder positioned at the origins of nineteenth-century program music, and whose strange and novel output is considered at some length in Grove's first *Dictionary*. Sampieri was a castrato who transferred to London around 1780 to sing at the King's Theatre. His career stalled the following year—he was apparently incapable of singing in tune—and he devoted himself to teaching the fortepiano and performing and publishing his own music, composed according to "an entirely new plan" (of which more below). Figure 3.1 reproduces an engraving after a portrait by the London-based painter Charles Hayter. It shows Sampieri—portly, mild-featured, respectable of dress, and ostentatious of cravat—in the act of composing, with one quill pen poised above a book of manuscript paper, another resting on the shelf behind him. To call Sampieri a figure of minor importance is an overstatement: he was not only expunged from "Programme Music," but merits not a single mention in the entire *New Grove*.

Admittedly, even Corder presented Sampieri as a mere archival curiosity: "Mr. Julian Marshall possesses a number of compositions"—he confided, in the parochial idiom that characterized much of the first *Grove*—"of an obscure but original-minded composer of this time . . . , Signor Sampieri." Marshall later sold his collection to the British Library, where the sole extant copies of most of Sampieri's works are gathered in a series of leather-bound volumes.[4] A brief survey of titles will suffice to give a sense of Sampieri's ambitions: there is a *Novel, Sublime, and Celestial Piece of Music, called*

Figure 3.1. Nicola Sampieri, by Charles Hayter, engraved by Thomas Turnbull.
(Music Division, The New York Public Library for the Performing Arts,
Astor, Lenox, and Tilden Foundations.)

Night (ca. 1800); a few years later, *Four Pieces of Music in Imitation of the Four Seasons of the Year. With four analogous and most elegant engravings* (the British Library dates this to 1806). The years around 1810 saw *The Various Motions of the Sea*, as well as *The Magic Lantern* and *The Progress of Nature in Various Departments*. Around 1815 he published his only programmatic duet, *A Grand Miscellaneous, Curious & Comical Piece of Music called The Fair*, for violin and piano. This included (among other scenes) a dancing bear, dancing dogs, a woman shaving a man, and a boy on a swing; it finished with "the signing of THE DEFINITIVE TREATY and the Restoration of France, or Peace Proclaimed between Two Great Nations, Announced by the Sound of Trumpets, Bells ringing, playing merry Peals, & the Firing of the Guns."

A sense of Sampieri's compositional style may be gleaned from example 3.1, which reproduces four brief excerpts from *The Various Motions of the Sea*. Example 3.1a depicts the sea in a state of "calm," and then "perfect calm";

Perfect Calm

Example 3.1a. Nicola Sampieri, excerpt from *The Various Motions of the Sea*

the hands in unison slowly trace the lower tetrachord of G major, and a fleeting turn to chordal texture at the end allows Sampieri to match a cadence to "perfect calm." (The passage also reveals him as less than a stickler with regard to the number of parts, or their voice-leading.) In example 3.1b, a little later in the piece, "the waves begin to rise." Sampieri depicts the shape of the waves by means of arpeggios that rise and then reset, the figure moving sequentially up the scale. Finally the waves become "extremely high," and "extremely agitated"; in example 3.1c, Sampieri depicts the height of the waves by moving the right hand up to *e-flat'''*; the left hand oscillates within an E-flat major chord. Extreme agitation may be observed in example 3.1d, where frantic octave patterns in the right hand alternate with a thudding bass line in the left.

These excerpts are representative of the piece as a whole: *The Various Motions of the Sea* consists of wave patterns of various heights and lengths, represented by undulating scales and arpeggios, essentially in a raw state. There is little melody, apart from a sea shanty quoted at the opening; harmonic motion is of the simplest, Sampieri usually achieving it by means of wave patterns in rising or falling sequences. In short, it is not difficult to see why a present-day historian concerned with musical beauty should pass over *The Various Motions* in silence. It is instrumental music bound, in the most extreme way, to the depiction of visual phenomena. What is more, Sampieri's extra-musical objects are—at least at first glance—no more original than his melodies. It was hardly a novel idea in 1806 to represent the seasons in music: if Vivaldi's concerti were not fresh in listeners' minds, then Haydn's oratorio certainly was. And the lengthy military sequence at the end of *The Fair* drew on a tradition of battle pieces for piano that had become particularly robust during the Revolution and Napoleonic Wars.

It would seem, then, that Nicola Sampieri was no more distinguished as a composer than he had been as a singer. This itself says something: his contract for the 1781–1782 season at the King's Theatre engaged him "only until another castrato could be found," and he had to sue for payment.[5] Susannah Burney's diary entry on Sampieri's performance at a variety concert reminds that faint praise can be damning indeed: "even Sampieri I could bear as there are such pretty passages in his songs, though they are of a 2nd or perhaps 3rd rate."[6] The composer's lack of (a certain kind of) musical imagination did not go unnoticed by contemporaries. In 1816, one reviewer suggested creating "a scale of musical excellence from Beethoven down to Sampieri."[7]

Are there worthy motives for rescuing him from the obscurity into which he has fallen—motives, that is, other than a spirit of cheery antiquarianism, or a delight in musical oddities and failures? I will supply a handful of details in his defense, to set Sampieri apart from myriad other composers who produced now-forgotten fodder for amateurs in London during these years. Item one (see figure 3.2) is a playbill for a recital given by Sampieri, for what he called a "CONCERT upon an Entirely NEW PLAN."[8] The advertisement did not mislead: this "new plan" entailed an evening devoted exclusively to Sampieri's music (the single-composer recital was a novelty during this period).

Example 3.1b. Nicola Sampieri, excerpt from *The Various Motions of the Sea*

Example 3.1c. Nicola Sampieri, excerpt from *The Various Motions of the Sea*

Example 3.1d. Nicola Sampieri, excerpt from *The Various Motions of the Sea*

Even more unusual: as he played, a group of assistants oversaw the projection of analogous images by means of magic lanterns, somewhat in the manner of the phantasmagoria show although without its predilection for gothic thrills.[9] Thus, for example, Sampieri's "extremely agitated" musical waves would sound just as a transparent image of an agitated seascape was inserted into a machine equipped with a powerful light source. These images were supplemented by rattling sounds made by metal sheets, by gunpowder, resin, and strategic bursts of fire.

Sampieri seems to have begun these concerts in the late 1790s; in London, as figure 3.2 attests, they took place in 1798 in the premiere concert hall, the Hanover Square Rooms. He also took his "new plan" to other musical centers of the British Isles, being greeted with mixed success. In Chichester, a mere fourteen people attended (and after the event one disappointed patron returned his copies of Sampieri's music to the composer in lieu of payment, claiming that he "sho'd not play them if he had them"). The desired effect may have been greater in Leicester, where a chronicler enthusiastically recommended several decades later that "music-hall caterers might profitably seek a spiritualistic *séance* with Signor Sampieri."[10] He was still giving these signature concerts as late as 1808, when he was seen and heard in Dublin playing alongside "numerous and beautiful transparencies."[11]

Item number two in Sampieri's defense: he ensured that these analogies between the musical and the visual were preserved when his program music was published for domestic use. All of the pieces listed above were illustrated with "analogous" or "descriptive" engravings and contained precise instructions for matching details in the images to their corresponding music. For the most part, these engravings were placed in the top third of the page, above the music that represented them. For instance, the title page for Sampieri's *Night* enumerated the following plates, substituted for the transparencies in his concert:

> 1st Plate. Evening, representing Jupiter, Venus, and other Stars
> 2nd Midnight, The Moon gradually rising
> 3rd and 4th Aurora & Daylight, The break & encrease of Day, with the
>   Notes of various Birds
> 5th the Rising of the Sun in full Splendour.[12]

On the back leaf Sampieri supplied an additional "Short Account how this Piece is to be Expressed." The instructions ascribe a few basic affects ("cheerful," "serious," "innocent," "animating") to what he called the "most stupendous and wonderful change of the atmosphere," as well as revealing some confusion concerning the structure of the solar system:

> As it is supposed the Day is more Chearful than the Night, in consequence
> of which, the *Evening*, begins by a Piece of Serious Music. *Midnight*, by
> simple and innocent, at the same time shewing the Horror & Dead of the

HANOVER-SQUARE ROOMS.

CONCERT upon an entirely NEW PLAN.

SIGNOR SAMPIERI

Begs Leave to inform the Nobility, Gentry, and the Connoisseurs of Music in general, that

On MONDAY, the 21st of MAY, 1798,

Will be exhibited an extraordinary

MUSICAL PERFORMANCE,

OF HIS OWN INVENTION AND COMPOSITION.

ACT I.

A Piece of Music called NIGHT,

Divided into Five Parts, viz.

EVENING, MIDNIGHT, AURORA, DAYLIGHT, and the RISING of the SUN;

And in order to gratify the Audience more highly, and add grandeur to the Effect of the Music, the Room will be decorated with TRANSPARENCIES, in the following Manner:

EVENING will be displayed by a Transparency of *Jupiter, Venus,* and other *Stars.*

MIDNIGHT by a Transparency, with a View of the *Moon* gradually rising.

AURORA and DAYLIGHT by the Notes of the various Birds, in a surprising Manner.

The Stars and the Moon will then gradually disappear, and Daylight be followed with the Sun rising in splendour.

The Second Piece, Four Pieces of Music in Imitation of

The FOUR SEASONS of the YEAR.

And the Third Piece of Music,

A SERENATA,

With a Description in the Bill at the Room.

ACT II.

A Piece of Music, called

THE SEA STORM,

With the SHIPWRECK and a Prospect of the SEA in MOTION.

The next is a Piece of Music, Vocal and Instrumental, called

THE JUDGMENT OF PLUTO,

With the Trial of the Prisoners and their Sentence, producing a grand and awful Effect.

To conclude with

A Piece of MUSIC, representing the Punishment to the condemned by the INFERNAL REGION.

The last Piece of this Grand Entertainment is called

LE ROY A LA CHASSE,

Interrupted by a THUNDER STORM;

the Course of the Storm, the Music will express the Agitation of the Wind, the Roaring of the Thunder, the Flashes of Lightning, and the Falling of Thunder Bolts, till the whole Land shews that of a Deluge; after which, comes a gentle Calm.

At the End of the above will be introduced a favourite HUNTING SONG, and conclude with a CHORUS.

In the Course of the Concert will be introduced

OVERTURES, SONGS, and other favourite PIECES.

After the Concert (by particular Desire of several Ladies and Gentlemen) will be

A BALL.

B. The Music is entirely New and analogous to the above Pieces. Particulars in the Bill of the Concert.

TICKETS, 10s. 6d. each, to be had at the Rooms.
To begin precisely at Eight o'Clock, and the Ball at Eleven.

Figure 3.2. A Playbill for Sampieri's "Concert on an Entirely New Plan."

Night. *Aurora*, by a Mild encreasing [*sic*] swelling or crescendo Music, to shew the gradual approach of the Day. *Daylight*, by a Gay & pleasing Movement, *the Rising of the Sun*, concludes by an animating & lively Rondo, & as the Sun advance into the Centre of the Globe, the more the Music is animating, and finishes the piece.

Sampieri may be seen holding a score of his signature design in the portrait reproduced in figure 3.3a, which was affixed to the front of his *Four Seasons of the Year*. The portrait, by an unknown artist, shows Sampieri's keyboard and quills before an expanse of landscape, implying that he was inspired by direct observation of nature itself as he composed. The illustrated scores in this design were "printed for the author," and sold privately at his house. Probably owing to the expense, Sampieri restricted himself to four to five engravings per piece before the 1810s. Thus each engraving usually had to stand for musical segments that had multiple objects. By the time of *The Fair*—that is, 1814 or a little later—he had hit on a new solution. Instead of illustrations at the beginning of each section, he included much smaller engravings between, above, and below the individual lines of music. In *The Fair*, interlinear illustrations depict a dancing bear, dancing dogs, and even a battle between the French and the English infantry and cavalry.

The military segment of *The Fair* would seem to stand in mysterious relation to its other carnival and pastoral offerings. Its connection to the nascent genre of program music, on the other hand, is less perplexing. Battle music accounted for the rest of the program music published in England during these years, virtually without exception.[13] Highlights include Koczwara's *Battle of Prague* (London, 1790), Steibelt's *Britannia: An Allegorical Overture in Commemoration of the Signal Naval Victory obtained by Admiral Duncan over the Dutch Fleet the 11th of October 1797* (London, 1797; dedicated to the king), and Ferdinand Kauer's *Wellington's and Blucher's Famous Battle near Waterloo* (London, 1815). The pianos themselves were often decorated with battle scenes and equipped with appropriate noisemakers.[14] During the Revolutionary and Napoleonic years, such pieces aided in stirring nationalist sentiment in favor of war. For instance, Jan Ladislav Dussek's symphony *The Sufferings of the Queen of France . . . Expressing the feelings of the unfortunate Marie Antoinette, during her imprisonment, trial, &c*, was adapted for the fortepiano or harpsichord and published in Edinburgh in 1793. *The Sufferings of the Queen of France* stimulated pro-Bourbon sympathies by depicting Marie Antoinette's anguish as she was separated from her children, her prayers before death (played *devotamente*), and even the fall of the guillotine.

Sampieri's aims, however, were different—and here we arrive at the

Figure 3.3a. Frontispiece to Nicola Sampieri, *Four Seasons of the Year*.
(Music Division, The New York Public Library for the Performing Arts,
Astor, Lenox, and Tilden Foundations.)

Figure 3.3b. Detail from frontispiece to Nicola Sampieri, *Four Seasons of the Year*.
(Music Division, The New York Public Library for the Performing Arts, Astor,
Lenox, and Tilden Foundations.)

most important item in our exhibit. His program music was written not for
drawing-room patriots but for "the lovers of science" (or, in the case of *Night*,
to "the ladies of science"). Figure 3.3b reproduces a detail from the title page
of Sampieri's *Four Seasons of the Year*: the dedication "To the lovers of science" more prominent even than the title, perhaps ambitiously so from a
composer who believed that morning arrived when "the sun advance[d] into
the centre of the globe." What is more, while composers often sought an
audience in their dedications, "the lovers of science" were not (until then)
considered a significant portion of the sheet-music market—at least, not
compared to more frequent dedicatees such as "London's lady amateurs" or
"young musicians."

Why, then, did Sampieri claim a scientific audience? A clue can be found
in his insistence on the principle of analogy: the plates in his scores were
not mere illustrations, but "analogous images." He was, in other words, careful to avoid setting music and image in hierarchical relation. He did not aim
to supply mere inspiration in the form of natural scenes, nor to recreate the
experience of the magic-lantern show, which often featured mechanical and
dance-like barrel-organ accompaniment alongside the main visual event.
Rather, Sampieri encouraged customers to experience the direct correspondence between shapes heard and shapes seen.

PARTAKING OF THE SAME MOTIONS

This dedication may direct our attention to what is, at the very least, a striking historical coincidence. Sampieri's "New Plan" for a pictorial music came
into being at the precise moment, and in the very city, in which institutional-

ized science was radically recalibrating its accounts of how music and images were transmitted and perceived, in the process creating its own "analogy of light and sound." This recalibration began with a paper read at the Royal Society in 1799 by the young Quaker physician Thomas Young, published the following year as "Outlines of Experiments and Inquiries Respecting Sound and Light" (hereafter, "Respecting Sound and Light").[15] Young followed this with several more publications during the next decade, refining, elaborating on, and supplying new evidence for a single, groundbreaking hypothesis. The argument can be summarized simply enough: light, he argued, was not a particle, as prevailing wisdom had long held; rather, it was a wave, like sound.[16] According to this new understanding, individual sights and sounds traversed their own media—ether and air respectively—by means of a single mechanism, the undulation.

Did this new scientific "analogy of light and sound"—so Young labeled it—offer anything new to the ambitious craftsmen of sound–light analogies within the fine arts? The question requires us to consider in a little more detail both Young's analogy and the model it supplanted. According to subscribers to the corpuscle model, light comprised rigid, weightless particles that traveled rapidly from luminous bodies to the eye in straight lines. For Newton (and his followers), light could obviously not be a wave, like sound; if it were, then it should travel around corners, not merely in straight lines.[17] Newton had suggested that color corresponded to the velocity of the corpuscles as they encountered the eye; the prism separated white light into streams of different speeds. Though he believed that light was matter and sound was motion, Newton nonetheless suggested an analogy between the seven colors of his white light spectrum and the seven notes of the Pythagorean monochord. His *Opticks* (1704) presented the famous "color circle," which matched the notes of the scale to colors, throwing in the planets for good measure.[18]

Newton's analogical thinking impressed Young, but the corpuscles did not. Young aimed to identify a much more profound likeness between light and sound, one that encompassed medium and behavior rather than a simple coincidence of number. "Respecting Sound and Light" put forward an ostensibly haphazard assembly of observations, related to specific instrument technologies and freak acoustic phenomena.[19] One section offered a new understanding of the mechanism of organ pipes, another some observations about tuning systems. He declared that the technology of the speaking trumpet provided evidence against the emanation of sound waves equally in all directions.[20] Elsewhere, he observed that the human voice typically produced at least four overtones, while the strings of the violin, if struck in the middle, made "either no sound at all, or a very obscure one."[21] He then turned his attention to acoustic beats and so-called third, or "Tartini," tones.

Diverse as these musical objects may have seemed, Young brought them together to demonstrate a single new principle: wave interference. Sound waves that cross paths do not merely dodge each other's particles of air, as had previously been assumed. Rather they briefly coalesce, "each particle [of air] partaking of the same motions," with their original trajectories undisturbed.[22] The principle of interference explained why, for instance, the frequency of acoustic beats was equal to the difference of frequencies between the two tones; it was also why a cord struck at the aliquot did not make a sound. Young then used the principle of interference to explain a number of optical phenomena. These included Newton's Rings, an interference pattern created when light is reflected between a spherical surface and a flat surface; also the colors of thin plates, and diffraction fringes. In late 1802 he invented what he called a "harmonic slider" (see figure 3.4), which allowed the amateur to study wave interference in the comfort of her drawing room.[23] As befit a principle with so many applications, Young promised multiple uses for his slider, ones relating not merely to optics but to the calculation of musical consonance and dissonance and even to the patterns of sea tides.

These projects make clear why Young has been described as a proponent of a "unified physics"—a term most often associated with his older French contemporary Pierre-Simon Laplace.[24] But there are important distinctions between Parisian unified physics and its London counterpart. Laplace and his followers aimed to unite all domains of physical science through meticulous study of interparticle forces of attraction and repulsion: their baseline was of matter and, for this, light must be molecular. Young's, on the contrary, was a principle of medium: concerning the impulses or qualities that traversed it, and the organs that received them. Not for nothing did Laplace vehemently denounce Young's light waves; indeed, Young's experiments are numbered among the death knells for Laplacian physics. In the London model, undulations functioned as something like a *lingua franca*, uniting the "vernaculars" of the individual senses. While Laplace sought to establish his theory of a unified physics by means of unprecedentedly meticulous mathematics, Young disapproved of scientists who communicated their findings primarily through algebra. He did not eschew mathematics as a means, but he compared the scientist mired in calculations to a night traveler incapable of observing the scenery. Far more effective was analogy: Young suggested that analogy was "a most satisfactory ground of physical inference," even claiming that "this combination of experimental with analogical arguments, constitutes the principal merit of modern philosophy."[25]

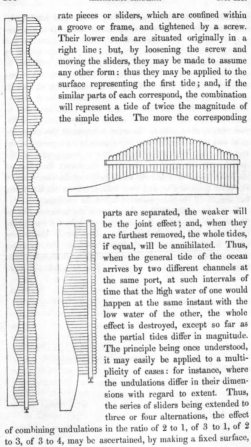

rate pieces or sliders, which are confined within a groove or frame, and tightened by a screw. Their lower ends are situated originally in a right line; but, by loosening the screw and moving the sliders, they may be made to assume any other form: thus they may be applied to the surface representing the first tide; and, if the similar parts of each correspond, the combination will represent a tide of twice the magnitude of the simple tides. The more the corresponding

parts are separated, the weaker will be the joint effect; and, when they are furthest removed, the whole tides, if equal, will be annihilated. Thus, when the general tide of the ocean arrives by two different channels at the same port, at such intervals of time that the high water of one would happen at the same instant with the low water of the other, the whole effect is destroyed, except so far as the partial tides differ in magnitude. The principle being once understood, it may easily be applied to a multiplicity of cases: for instance, where the undulations differ in their dimensions with regard to extent. Thus, the series of sliders being extended to three or four alternations, the effect of combining undulations in the ratio of 2 to 1, of 3 to 1, of 2 to 3, of 3 to 4, may be ascertained, by making a fixed surface,

Figure 3.4a–b. Wave interference modeled by Dr. Young's Harmonic Sliders. (Courtesy of the Thomas Fisher Rare Book Library, University of Toronto.)

junction of undulations, is rendered visible and intelligible, with great ease, in the most complicated cases. It is unnecessary to explain here, how accurately both the situations and motions of the particles of air, in sound, may be represented by the ordinates of the curve at different points; it is sufficient to consider them as merely indicating the height of the water constituting a tide, or a wave of any kind, which exists at once in its whole extent, and of which each point passes also in succession through any given place of observation. We have then to examine what will be the effect of two tides, produced by different causes, when united. In order to represent this effect, we must add to the elevations or depressions in consequence of the first tide, the elevations or depressions in consequence of the second, and subtract them when they counteract the effect of the first: or we may add the whole height of the second above any given point or line, and then subtract, from all the sums, the distance of the point assumed below the medium.

To do this mechanically is the object of the harmonic sliders. The surface of the first tide is represented by the curvilinear termination of a single board. The second tide is also represented by the termination of another surface; but, in order that the height at each point may be added to the height of the first tide, the surface is cut transversely into a number of sepa-

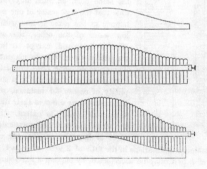

THE IMMEDIATE SOMEWHAT BETWEEN

It should be admitted at this juncture that there is no further evidence to bind together the musical and the scientific: no documentary trail to place Sampieri in Young's audience at the Royal Society or Young at the "Concert upon an Entirely New Plan." One well-trafficked bridge over such lacunae is the rhetoric of the paradigm shift. Along these lines, one might suggest that Young's new "analogy of light and sound" lay quietly behind Sampieri's project of designing visual–musical analogies for the concert hall and the drawing room. This kind of interpretive move has become common during recent decades of scholarship on literary romanticism and the history of science. Frederick Burwick, Trevor Levere, Mark Lussier, Alan Richardson, Sophie Thomas, and many others have demonstrated the extent to which the English romantic poets were informed by developments in "natural philosophy" and the attendant technologies for its observation.[26] Indeed, the extent and influence of this work are such that studies like the present one need no longer expend their energies deconstructing outdated binaries between romantic transcendence and scientific materialism.

These scholars have shown Young's theories on light waves to be particularly influential to contemporary poets. Lussier, for instance, has suggested that Shelley studied Young's publications and based his model of "poetic complementarity" on the understanding of wave dynamics thereby acquired.[27] (Did he also purchase the harmonic sliders?) But the aesthetic influence of Young's work stretched well beyond Shelley's poetics; indeed, it can be detected in some of the most famous sound-light analogies in English romantic texts. For instance, Wordsworth opened his 1828 poem "On the Power of Sound" by first addressing the eye, and then describing the ear:

Thy functions are ethereal,
As if within thee dwelt a glancing mind,
Organ of vision! And a Spirit aerial
Informs the cell of Hearing, dark and blind.[28]

These lines grapple with a tension between immateriality of medium—"ethereal" for the eye, "aerial" for the ear—and fleshly substance and shape. Perhaps most relevant here, though, is the pun on "glancing" in the second line, which evokes both an intelligence that sees and one that passes over, glides past, reflects off—darting through the spaces between matter, as light itself does. Coleridge surely had Young's undulations in mind when, in 1817, he interpolated four lines into his "Eolian Harp." The revised poem extolled:

the one life within us and abroad,
which meets all motion and becomes its soul,
A light in sound, a sound-like power in light
Rhythm in all thought, and joyance everywhere.[29]

In both of these canonic poems, the equation of light's properties with those of sound served to index an all-encompassing natural order.

Was Sampieri's project also a harbinger of this order? In the absence of evidence binding his music-and-light shows more tightly to the mast of hard science, we might subsume pieces like *Night* and *The Progress of Nature in Various Departments* within a broader multidisciplinary discourse facilitated by the "mediumnist" conception of light waves. Certainly new analogies between sound and light seem to have become possible around 1800: ones that (as we have seen) relied on a conception of both as coextensive media manipulated by a single mechanism. This rapprochement in turn facilitated a decisive break with earlier understandings of a fundamental separation between the arts of sound and the arts of matter. To take the paradigmatic instance of this earlier line of thought, Lessing had written in his *Laokoon* that visual art extended in space, while poetry and music extended through time—an understanding that aligns comfortably with the distinction between sonic undulations and luminous corpuscles. For Lessing, even analogical turns of phrase like *poetische Malerei* betrayed an ignorance of this essential difference between matter and motion.[30]

When light came to be understood as motion-in-medium rather than as material, it could be explained with sound as the point of reference, following the same principles of movement. As William Phillips summarized in 1817, "Light is . . . only a quality or influence, conveyed to the eye by means of the vibratory motions belonging to the medium which conveys it; just as sound is conveyed to the ear, by the motions of our atmosphere."[31] Such figures were easily adapted to theoretical formulations about the relation of the fine arts to each other. For instance, the English composer William Gardiner suggested that the province of music, like that of poetry and painting, is to "represent a picture of nature"; and thus "it may be said of music, as Coleridge said of painting, that it is the immediate somewhat between a thought and a thing."[32] Notable in Coleridge's figure, and in Gardiner's repurposing, is the emphasis on somewhat-ness and insubstantiality: paintings, and by extension music, become neither object, nor idea, but rather the space that separates the two.

There is some ground for thinking of Sampieri's program music as a new turn toward the audiovisual in program music—and I use this anachronistic language on purpose, to invoke a mode in which sights and sounds are brought into being by shared submaterial, subperceptible patterns. Sampieri's music simply mimicked the changes of "atmospheric" light and the contours of its motion. His musical waves could thus participate in what Sophie Thomas has described as a "formative dialogue between the media and the material, the theorization and the fabrication of the visible."[33] Since the wave model only requires that music goes up and down, this hermeneutic move might seem deflatingly facile. But what might bear weight here is not simply the regularity of the rises and falls but the fundamental nature of Sampieri's materials, their nature *as* pure fundaments. His unceasing undulations through scales and arpeggios, and his exclusive reliance on sequential motion and parataxis, need not imply simply an inept melodic imagination. Rather, they could reflect the delights of a different kind of imagination: one that drew new kinds of connections between music, visions, and the natural order, and for which science seemed briefly poised to lend its assistance. In this sense (admittedly a fanciful one), Sampieri's visual music is not the Pastoral Symphony's idiot cousin, but a worthy successor to the first melodrama scores, and precursor of minimalist music such as *Einstein on the Beach*: in each of these cases, oscillating scales and arpeggios come to the fore as music is bespoke to a new visual medium. From this angle, Scruton's notion that programs force music to "forsake its own autonomous principles" is misleading at best: what are more autonomously musical than scales and arpeggios? And yet these are rarely more prominent than when music gives itself over to visual depiction.

However, a triumphalist account of mediumnist sound–light analogies — a summoning of trumpets and drums to drown out the "third-rate" strains of an aging castrato at the fortepiano, and the weak applause of all fourteen of his spectators—leaves this historian, at least, with a bad conscience. There is a difference between drawing on the history of science to annotate Coleridge's "Eolian Harp" and using it to confer new presence on artworks that contemporaries made no attempt to preserve. Distinctions of quality were crucial to aesthetic experience in the age of Coleridge, Young, Wordsworth, and Sampieri, and very similar ones continue to be made today. Certainly, the conventions of post-canonical humanism supply something of a get-out-of-jail-free card here, allowing serious scholarship about a kind of music that is (by any familiar standard) simplistic, inept, and malformed. But is it possible, instead, to confront the issue of quality head on?

What kind of aesthetic attention did Sampieri's music request from its listeners? What kind of listening did it reward? To put this differently, is there a way in which we can take Sampieri's music seriously? If there is, it should start by acknowledging the music's considerable deficiencies. The problem with music like Sampieri's is not that it yields nothing to the tools of harmonic analysis; rather, it yields pretty much everything. To paraphrase Nicholas Mathew, music needs to resist analysis at least a little for analysis to find it interesting.[34] There is no dichotomy to be drawn in Sampieri's music between the essential and the inessential, structure and surface, affective and artful. Permeated with objects, Sampieri's music is devoid of subject; it is just motion, just material, just physics; you can see right through it. As such, it occupies the same status as the color-painted slide passed through the aperture of the lantern as he played. Neither the slide nor the music gives rise to discourses of permanence, reification, cherishing, repeated experience; neither asks to be kept. Rather, they rewarded indexical response, pointing, they trained in object classification and analogic equivalences: *that* is a wave, *that* is a sunrise, in music. And as such, the value of uniqueness of the art-object cedes to one of exemplarity; the exemplar is valuable only insofar as it facilitates easy swinging between branches on the great tree of natural philosophy.

This notion of Sampieri's music as educative in a broader indexical listening may well give pause, again precisely because of the raw nature of his musical indices. If music's ups and downs are permitted to signify spatially or materially—if a rising scale cues a moonrise, and undulating arpeggios immediately toss up ocean waves before the mind's eye—what is to prevent such banalities from intruding on all musical experience? Whether the listeners who think of a storm at sea when they hear a Mozart sonata are engaging in a socially valuable form of analogic thinking, or merely displaying a dangerously susceptibility to imposing "naïve pictorialism," is of course a loaded question both historically and socially. As we have seen, this was the node of disagreement between Scruton's and Corder's accounts of the history and value of program music, and between Adorno's and Disney's notions of the "cartoon" or "cartoonish" musical experience. If a broader history concerning the moral implications of tone-painting must remain outside the scope of this paper, a point of continuity can at least be noted between Sampieri's pieces of program music and Frederick Corder's definition thereof, sketched in the same city eighty years later. Corder had the terms at his disposal to confer social value, if not praise of execution, on Sampieri's transparent musical indices, and he sketched a musical canon on that basis. These terms now sit squarely alongside Sterndale Bennett's *Paradise and the Peri* Overture in the dustbin reserved for artifacts of *das Land ohne Musik*, immediately identifiable according to their distinctive blend of the risibly earnest, the banal, and the parochial. Yet the value assigned to analogic thinking and listening in both Sampieri's and Sir George Grove's London calls for renewed attention to the history of musical significance in that nineteenth-century metropolis: terrain that has become, in its very historical proximity, so profoundly distant.

NOTES

1. Roger Scruton, "Programme Music," *Grove Music Online, Oxford Music Online,* accessed October 12, 2013, http://www.oxfordmusiconline.com/subscriber/article/grove/music/22394.

2. Quoted in Esther Leslie, "Eye Candy and Adorno's Silly Symphonies," in *Hollywood Flatlands: Animation, Critical Theory and the Avant-Garde* (London: Verso, 2002), 160.

3. F. C. [Frederick Corder], "Programme-Music," in *A Dictionary of Music and Musicians (A.D. 1450–1889) by Eminent Writers, English and Foreign. With Illustrations and Woodcuts,* ed. Sir George Grove (Philadelphia: Presser, 1895), 3:34–40 (emphasis added).

4. See Arthur Searle, "Julian Marshall and the British Museum: Music Collecting in the Later Nineteenth Century," *British Library Journal* 11 (1985): 67–87.

5. Judith Milhous and Robert D. Hume, "Opera Salaries in Eighteenth-Century London," *Journal of the American Musicological Society* 46, no. 1 (Spring 1993): 45.

6. Philip Olleson, ed., *The Journals and Letters of Susan Burney: Music and Society in Late Eighteenth-Century England* (Surrey, UK: Ashgate, 2012), 150–51. Burney also noted that Sampieri's singing was "insufferably out of tune."

7. "Review of New Musical Publications," *Gentleman's Magazine and Historical Chronicle,* January 1816, 60.

8. The playbill in figure 3.2 is reproduced in M. Phillips and W. S. Tomkinson, *English Women in Life & Letters* (Milford: Oxford University Press, 1927), 240. On the concert space in the Hanover Square Rooms, see Michael Forsyth, *Buildings for Music: The Architect, the Musician, and the Listener from the Seventeenth Century to the Present Day* (Cambridge: Cambridge University Press, 1985), 35–39.

9. See Mervyn Heard, *Phantasmagoria: The Secret Life of the Magic Lantern* (Hastings: Projection Box, 2006). An exception may be found in the representation of "Midnight" in Sampieri's *Night,* which aimed to evoke "the Horror & Dead of the Night"; see below.

10. On the Chichester concerts, see Brian Robins, ed., *The John Marsh Journals: The Life and Times of a Gentleman Composer (1752–1828)* (Hillsdale, NY: Pendragon, 1998), 582; on those in Leicester, see Robert Read, *Modern Leicester: Jottings of Personal Experience and Research* (London: Simpkin, Marshall, & Co., 1881), 212. Marsh wrote of the Chichester concerts that "for the finishing piece each time Sampieri had a symphony expressive of a storm in which was at particular parts, an accompan't of thunder & lightning behind a screen, the former of which was done by shaking a large piece of tin & the latter by a little powder of rosen blown thro' a bit of paper into the flame of a candle" (212).

11. John C. Greene, ed., *Theatre in Dublin, 1745–1820: A Calendar of Performances* (Plymouth: Lehigh University Press, 2011), 3616–17.

12. This table of contents recalls those given in the programs for the eidophusikon; see Deirdre Loughridge's essay in this volume.

13. For a catalogue of eighteenth-century orchestral works with programs, see Richard Will, *The Characteristic Symphony in the Age of Haydn and Beethoven* (Cambridge: Cambridge University Press, 2001), 249–303.

14. Arthur Loesser, *Men, Women, and Pianos: A Social History* (New York: Simon and Schuster, 1954), 166–74.

15. Thomas Young, "Outlines of Experiments and Inquiries Respecting Sound and Light . . . In a letter to Edward Whitaker Gray," *Philosophical Transactions of the Royal Society of London* 90 (1800): 106–50; quotation at 146. On Young's career, in addition to the works cited below, see George Peacock, *Life of Thomas Young* (London: J. Murray, 1855); Alex Wood and Frank Oldham, *Thomas Young, Natural Philosopher, 1773–1829* (Cambridge: Cambridge University Press, 1954); Jed Z. Buchwald, *The Rise of the Wave Theory of Light: Optical Theory and Experiment in the Early Nineteenth Century* (Chicago: University of Chicago Press, 1989); and Andrew Robinson, *The Last Man Who Knew Everything: Thomas Young, the Anonymous Polymath Who Proved Newton Wrong, Explained How We See, Cured the Sick and Deciphered the Rosetta Stone* (New York: Pi, 2007).

16. Young was not the first to suggest a wave model for understanding light, but he was more successful than earlier wave theorists, such as Christiaan Huygens and Leonard Euler, in dismantling major objections to the idea. What was more, he did so before an English audience, for whom the adherence to Newton's corpuscular model of light was a point of national pride.

17. Newton, *An Hypothesis Outlining the Properties of Light* (1675); quoted in Oliver Darrigol, *A History of Optics from Greek Antiquity to the Nineteenth Century* (Oxford: Oxford University Press, 2012), 89–90.

18. See Emily Dolan, *The Orchestral Revolution: Haydn and the Technologies of Timbre* (Cambridge: Cambridge University Press, 2013), 23–31.

19. Peter Pesic, "Thomas Young's Musical Optics: Translating Sound into Light," in *Music and Sound in the Laboratory*, ed. Alexandra Hui, Julia Kursell, and Myles W. Jackson (Chicago: University of Chicago Press, 2013), 15–39.

20. Young, "Respecting Sound and Light," Section 4: "Of the Divergence of Sound," 118–20. Here Young cited Johann Heinrich Lambert's insights in his Berlin memoirs (1763), which Darrigol suggested he knew second-hand through John Robison's entry, "The Speaking Trumpet," in the *Encyclopaedia Britannica*, 3rd edition (1797), 18:583–93. See Darrigol, *A History of Optics*, 168.

21. Young, "Respecting Sound and Light," 139.

22. Here Young referred to Smith's *Harmonics, or The Philosophy of Musical Sounds* (London: Merrill, 1759): "Respecting Sound and Light," 130–31.

23. [Thomas Young], "An Account of Dr. Young's Harmonic Sliders," *Journal of the Royal Institution of Great Britain* 1 (1802): 261.

24. See, for instance, Robert Fox, "Laplace and the Physics of Short-Range Forces," in *The Oxford Handbook of the History of Physics*, ed. Jed Z. Buchwald and Fox (Oxford: Oxford University Press, 2013), 406–31.

25. Thomas Young, *A Course of Lectures on Natural Philosophy and the Mechanical Arts*, rev. ed. (London: Taylor and Walton, 1845), 1:5.

26. Frederick Burwick, *The Damnation of Newton: Goethe's Color Theory and Romantic Perception* (New York: De Gruyter, 1986), esp. 176–274; the essays in Andrew Cunningham and Nicholas Jardine, ed., *Romanticism and the Sciences* (Cambridge: Cambridge University Press, 1990); Trevor Levere, *Poetry Realized in Nature: Samuel Taylor Coleridge*

and Early Nineteenth-Century Science (Cambridge: Cambridge University Press, 2002); Mark Lussier, *Romantic Dynamics: The Poetics of Physicality* (London: Macmillan, 2000); Alan Richardson, *British Romanticism and the Science of the Mind* (Cambridge: Cambridge University Press, 2001); see also Mark Lussier, "Science and Poetry," in *The Encyclopedia of Romantic Literature*, ed. Frederick Burwick (Malden, MA: Blackwell, 2012). On the English Romantics' engagement with optical technologies, see Sophie Thomas, *Romanticism and Visuality: Fragments, History, Spectacle* (New York: Routledge, 2007).

27. Lussier, "Shelley's Poetics, Wave Dynamics, and the Telling Rhythm of Complementarity," *Wordsworth Circle* 34, no. 2 (Spring 2003): 91–95.

28. See especially James Chandler, "The 'Power of Sound' and the Great Scheme of Things: Wordsworth Listens to Wordsworth," in *"Soundings of Things Done": The Poetry and Poetics of Sound in the Romantic Ear and Era*, ed. Susan J. Wolfson (Romantic Circles Praxis Electronic Editions, 2008), http://www.rc.umd.edu/praxis/soundings/chandler /chandler.html. Burwick read this passage as reflecting not merely the "commonplace" propagation analogy of Young, but also the matter of sound and light perception as described by John Gough (*The Damnation of Newton*, 191–96).

29. See Burwick, *The Damnation of Newton*, 156–57.

30. Lessing, *Laokoon, oder, Über die Grenzen der Malerei und Poesie* (1766); published as *Laocoön: An Essay on the Limits of Painting and Poetry*, trans. Edward Allen McCormick (Baltimore, MD: Johns Hopkins University Press, 1984). As Martin Meisel has shown, Lessing's manual was one of two theories of the fine arts taken seriously in England—the other was Reynolds' *Discourses*—and it remained a touchstone for certain arts institutions (such as the Royal Academy) well into the nineteenth century. Meisel, *Realizations: Narrative, Pictorial, and Theatrical Arts in Nineteenth-Century England* (Princeton, NJ: Princeton University Press, 1983), 17–20.

31. Phillips, *Eight Familiar Lectures on Astronomy* (London: Phillips, 1817), 225–26.

32. William Gardiner, *Music and Friends: Or, Pleasant Recollections of a Dilettante* (London: Longman, Orme, Brown, and Longman, 1838), 1:216, 221.

33. Thomas, *Romanticism and Visuality*, 7.

34. Nicholas Mathew, *Political Beethoven* (Cambridge: Cambridge University Press, 2013), 7–8.

CHAPTER FOUR

Charles Wheatstone: Musical Instrument Making, Natural Philosophy, and Acoustics in Early Nineteenth-Century London

MYLES W. JACKSON

Sir Charles Wheatstone (1802–1875) is best known among physicists, engineers, and historians of science and technology as the inventor of the stereoscope, a scientific instrument used to display objects in three dimensions, and the Wheatstone bridge, an electric circuit used to measure the current of an unknown electrical resistance. What is more, he is generally considered to be one of the leading figures of nineteenth-century telegraphy. However, such descriptions ignore his early engagement with and training in musical instrument making. This chapter sheds light on the kinds of material objects with which experimental natural philosophers worked before laboratories such as Cambridge University's Cavendish existed. It also reminds us that, at least since the eighteenth century, musical instruments have been used for experimental inspiration: like the submarine cable in the 1850s and '60s, musical instruments and their modes of vibration provided additional resources and techniques for the nineteenth-century physicist.

Charles Wheatstone was born in 1802 in Gloucester to William and Beata Wheatstone; his father and grandfather were skilled craftsmen. When Charles was four, the family moved to London, where his father built musical instruments, such as flutes and bassoons, and offered musical instruction on both instruments. They found in London a market for those skills that was like no other in Britain. On the death of their uncle in 1823, Charles and his brother William took over the musical instrument shop on the Strand. In 1829 they moved it to 20 Conduit Street, where Charles also lived until he married in 1847.[1] As a teenager, Charles had become fascinated with the intricate mechanisms of various scientific instruments; he was particularly enthusiastic about miniature automata playing musical instruments. As we shall see, Wheatstone's early interest in musical instruments and automata informed his later work on the science of sound.

If he ended his career as a scientist of great seriousness and renown, though, he began as something of a conjurer. "Charley Wheatstone's Clever Tricks" captivated the imagination of London audiences in 1822, when he was just twenty years old.[2] In the spring of that year, he opened Wheatstone's Musical Museum (open daily from noon until 5:00 p.m.) under the patronage of Princess Augusta Sophia, the second daughter of King George III and Charlotte of Mecklenburg-Strelitz.[3] It was initially situated in the Royal Opera Arcade on Pall Mall, but later moved to the Great Room on Spring Gardens, which was the site of numerous popular exhibitions of all kinds. Wheatstone initially charged one shilling and later raised the price to five shillings for a one-hour concert. An advertisement proclaimed:

> These entertainments, which consist of performances in the most superior style, alternately succeed each other, forming an hour's amusement, calculated to afford the highest gratification to the Musical World, and to interest the Public by the novelty of the means employed, and the beauty of the effects produced.[4]

The cleverest of the young Charlie Wheatstone's tricks may have been the "enchanted lyre," or acoucryptophone, which was first reported by a number of London journals in September 1821 and was initially housed in his father's shop on Pall Mall, remaining a popular attraction for two years.[5] This and his "Invisible Girl" exhibition are considered by Melissa Dickson and James Q. Davies in this volume. But they were by no means the only attractions at the Wheatstone's Musical Museum. Another was his Grand Central Diaphonic Orchestra, which "astonishingly augments in richness and power every variety of musical tones; among the instruments employed to exemplify this principle, are the Oedephone, (an equivalent of a band of wind instruments,) and Stodart's Compensation Grand Piano Forte"; together, these devices produced "magnificent Effects" in "an Instrumental Concert of singular beauty."[6] Another account spoke of "the beautiful experiments of Professor Wheatstone . . . by which four of Erard's harps play sweet but mysterious music without visible hands, as the sounds are conducted to them by rods from instruments played upon by performers who are placed several floors beneath the lecture-room."[7] These instruments all sounded without human players, or rather, with the players concealed from view. In other words, the principles Dickson identifies as central to the acoucryptophone eventually gave rise to an entire orchestra of playerless instruments in Wheatstone's Musical Museum.

LONDON'S SPACES OF PERFORMANCE

Wheatstone's early work on musical instruments—and his youthful enthusiasm for the technologies and principles of musical automata—can be situated firmly within the context of the musical and scientific spaces of public demonstration in the Strand.[8] During the late eighteenth and early nineteenth centuries, Londoners were titillated by numerous museums and exhibitions of mechanical contraptions, testimonies of mechanicians' skill in the period. Of particular interest to well-heeled audiences were automata, especially those of the renowned Swiss clockmaker Henri-Louis Jaquet-Droz, who opened his Spectacle Mécanique in the Great Room at 6 King Street, Covent Garden in 1776. Three of his *Meisterwerke* were featured: "one figure writes whatever is dictated to it, another draws and finishes in a masterly manner several curious designs; another plays divers Airs on the Harpsichorde."[9] Replicas of the three original pieces, owned by the Swiss clock and automaton maker Henri Maillardet, appeared before London crowds well into the nineteenth century.[10] James Cox, a goldsmith and maker of baubles for the wealthy, exhibited automata that reached sixteen feet in height, including a ten-foot-high peacock clock with clockwork-driven automata, in the Great Room of Spring Gardens (later the home of Wheatstone's Musical Museum). Cox's museum was all the rage in London in the 1770s.[11]

Perhaps the most famous instrument maker and builder of automata in London at the time was John Joseph Merlin. This eccentric Belgian-born artisan had moved to London in 1760 after a stint in Paris, where he had impressed members of the Académie des Sciences. At some time during the 1780s, Merlin opened the doors of what would become one of London's most widely discussed museums, Merlin's Mechanical Exhibition.[12] The museum continued to attract the curious after Merlin's death in May 1803.[13] His "scientific toys," particularly his automata, were coveted objects on the market as late as the 1830s.[14] As Simon Schaffer has shown, one of the visitors to Merlin's Mechanical Museum was the young Charles Babbage, a figure discussed by Gavin Williams elsewhere in this volume.[15] Babbage's mother took her son (later the eleventh Lucasian Professor of Mathematics at Trinity College, Cambridge) to the museum on Prince Street around 1800. There, he was much struck by various harpsichords, clocks, and mathematical instruments. But one object in particular captured his imagination: an automaton of a young female dancer. After Merlin's death, the dancer was bought by a rival performer and entrepreneur, Thomas Weeks, who had just opened his own museum in the Haymarket and was hoping to lure Merlin's clientele. While working on the plans for his difference and analytical engines, Babbage (ever the procrastinator) once again thought of the automaton and purchased it for £35 from Weeks's auction. He restored the rather run-down dancer and placed her on prominent display in his Marylebone salon.[16]

The most infamous automaton of the 1780s was a chess-playing Turk. Created by the Austro-Hungarian mechanician Wolfgang von Kempelen, this celebrated contraption toured Europe, astonishing the public, some of whom went to great lengths to query whether it was a real automaton or merely an example of trickery. The Turk made its London debut in 1783–1784 with a return trip in 1818–1821, this second time under the guidance of Johann Nepomuk Mälzel, the renowned mechanician and pirate of the metronome.[17] Kempelen's contraption and Mälzel's subsequent improvements were deceptions: although they both went to great lengths to reveal the so-called mechanism of wheels, cranks, and shafts behind the Turk, in reality a small chess player was hidden inside the cabinet.

By the end of the eighteenth century, London had become home to numerous entertainment enterprises, such as Henry Barker's Leicester Square Panorama, established in 1806. Barker later expanded parts of his show to the Strand. John Scott built his theater, the Sans Pareil, which debuted shows involving optical effects and magic lanterns, also in 1806.[18] One of the most famous theaters was Haddock's Androides (also referred to as the Mechanical Theater), in Norfolk Street, which opened in 1794.[19] Its pieces included the spelling automaton, the French telegraph, and the Highland oracle.[20] The Lyceum (also known as the English Opera House) on Wellington Street, just off the Strand, offered a venerable venue for the demonstrations of recent technological inventions. (In the present volume, Deirdre Loughridge considers the Walkers' eidouranion, a vast orrery exhibited at the Lyceum in tandem with a lecture on astronomy.) The Lyceum's commitment to technological advancement extended to its own theatrical lighting and stage mechanisms. For example, in 1817 it became the first theater in the world to use gas lighting. Three years later, James Robinson Planché's *The Vampire* introduced a new piece of machinery, the so-called vampire trap, now referred to as a trapdoor; in 1824 Carl Maria von Weber's *Der Freischütz* impressively displayed numerous special effects. But perhaps the best example of the Lyceum's dedication to the presentation of new, wondrous technologies was the demonstration of a primitive telegraph, which had been a part of Haddock's Androides exhibition:

Explanation of the TELEGRAPHE [*sic*], to be exhibited every Evening . . . .
The TELEGRAPHE is an instrument at present used in France, for the

conveyance of certain intelligence, at the rate of 200 miles an hour, and which is effected without the knowledge of any persons, except those at the two extreme distances. The Scene is supposed to represent the country between Lille and Paris; and to try the effects of the Machine, four distances are appointed, as sufficient to convey a true idea of the ingenuity and utility of the Telegraphe.[21]

By the 1830s, many of these mechanical devices had begun to shed their associations with entertainment and parlor magic; instead, they came to be seen as critical to industry, to the powering of Britain's economy and empire. Only occasionally described as deceptions and objects of wonder, these contraptions were now seen as instruments to explore and exploit natural phenomena. As the Scottish experimental natural philosopher David Brewster aptly summed it up in 1832:

> The passion for automatic exhibitions which characterised the eighteenth century gave rise to the most ingenious mechanical devices, and introduced among the higher order of artists habits of nice and accurate execution in the formation of the most delicate pieces of machinery. . . . Those wheels and pinions, which almost eluded our senses by their minuteness, reappeared in the stupendous mechanism of our spinning-machines, and our steam-engines. The elements of the tumbling puppet were revived in the chronometer, which now conducts our navy through the ocean; and the shapeless wheel which directed the hand of the drawing automaton has served in the present age to guide the movements of the tambouring engine. Those mechanical wonders which in one century enriched only the conjurer who used them, contributed in another to augment the wealth of the nation; and those automatic toys which once amused the vulgar, are now employed in extending the power and promoting the civilization of our species.[22]

In short, London in general, and the area around the Strand in particular, offered unique spaces for showmen where emerging disciplinary standards of productivity, industry, and serious knowledge were dependent first on a vibrant popular fascination for natural philosophy, magic, and commerce.[23]

WHEATSTONE'S EARLY WORK ON VIBRATIONS

Acoustical phenomena such as those exhibited at Wheatstone's Musical Museum simultaneously delighted the public and served as building blocks for a better understanding of natural principles. It would be a mistake, therefore, to argue that Wheatstone's "clever tricks" were merely an example of showmanship: the creation and public demonstrations of his numerous musical contraptions provided him with material for scientific investigations, and he made a number of important contributions to early nineteenth-century acoustics. As Aileen Fyfe and Bernard Lightman point out, "The skill of the showman, the writer, the lecturer, or the curator might be acknowledged but distinguished from scientific expertise by being 'merely' the expertise of the performer. Yet, there was no such clear dichotomy in the nineteenth century."[24] In contrast to a number of examples of the "popularization of science" of the period (like Adam Walker's astronomy lectures, considered by Deirdre Loughridge in this volume), Wheatstone's displays and showmanship preceded research on acoustics, rather than the other way round.

Wheatstone's earliest works in the physical sciences were devoted to the science of acoustics and sound vibrations, which were natural phenomena relevant both to the study of the propagation of sound and the construction of musical instruments. His first scientific paper owed much to the earlier work of the German acoustician E. F. F. Chladni.[25] However, the two men differed in one important respect. Whereas Wheatstone could ply his craft in London, there was no equivalent city for Chladni on the Continent. Indeed, the latter was a peripatetic scholar: he journeyed throughout the German territories, Holland, and Brussels, his tour culminating in an audience with Napoleon in Paris, where he gave demonstrations and attempted to sell his two musical inventions, the clavicylinder and euphone.

After replicating Chladni's most famous experiments—those generating the so-called Chladni figures—Wheatstone was convinced that there must be tiny vibrations that Chladni had not observed and that contributed to timbre. As he explained in an article published in 1823 and illustrated a decade later (see figure 4.1), Wheatstone reckoned that these small vibrations could be observed by placing water on a vibrating plate.

> I took a plate of glass capable of vibrating in several different modes, and covered it with a layer of water; on causing it to vibrate by the action of a bow, a beautiful reticulated surface of vibrating particles commenced at the centres of the vibrating parts, and increased in dimensions as the excursions were made larger. When a more acute sound was produced, the centres consequently became more numerous, and the number of coexisting vibrating particles likewise increased; but their magnitudes proportionably diminished.

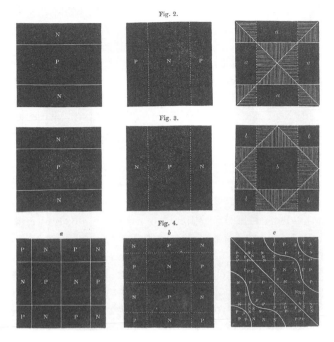

Figure 4.1. Chladni prints in Wheatstone's "On the Figures Obtained by Strewing Sand on Vibrating Surfaces, Commonly called Acoustic Figures," *Philosophical Transactions of the Royal Society of London* 123 (1833): 617. (Doe Library, University of California, Berkeley.)

He continued by explaining how Chladni figures are produced:

> The sounds of elastic laminæ are generally supposed to be owing to the entire oscillations of the simple parts, as shown by Chladni when, by strewing sand over the sonorous plates, he observed the particles repulsed by the vibrating parts accumulate on the nodal lines and indicate the bounds of the sensible oscillations.

However, Wheatstone surmised (quite correctly, as it turns out) that the vibrations Chladni revealed were not the only ones responsible for the generation of the sound.

> Did no other motions exist in the plate but these entire oscillations, the water laid on its surface would, on account of its cohesion to the glass, show no peculiar phenomena; but the appearances above described clearly demonstrate that the oscillating parts consist of a number of

vibrating particles of equal magnitudes, the excursions of which are greatest at the centres of vibration, and gradually become less as they recede
further from it, until they become almost null at the nodal lines.[26]

With the assistance of a micrometer, he proceeded to count the number of
vibrations generated by applying a violin bow perpendicularly to a metallic
plate covered with water, generating similar geometric shapes by blowing
through the open end of a flute or bassoon placed on the surface of a vessel
containing water.[27]

Vibrations also played a key role in Wheatstone's kaleidophone (or phonic
kaleidoscope), which was the subject of an important essay he published in
1827.[28] The "Description of the Kaleidophone" is significant for two reasons.
First, it acknowledged the interreliance of natural philosophy and public di-
version. In Wheatstone's words, "The application of the principles of science
to ornamental and amusing purposes contributes, in a great degree, to render
them extensively popular." He brought his kaleidophone to the attention of
London audiences because "it exemplifies an interesting series of natural
phenomena, and renders obvious to the common observer what has hitherto
been confined to the calculations of the mathematician." Secondly, the essay
sought to bind more closely together two fields of physics: acoustics and
optics. As he explained, the kaleidophone was the sonic equivalent of Brewster's kaleidoscope, invented in 1819. Thus the instrument "presents another
proof, that however remote from the common observation the operations
of nature may be, the most beautiful order and symmetry prevail through
all."[29] Wheatstone came up with the idea for his kaleidophone after reading
Thomas Young's experiments, as detailed in the *Philosophical Transactions*
of 1800: Young wound a silver wire around the lowest strings of a piano; the
silver reflected the light from a window, thus allowing Young to study the
wire's path once the keys were struck.

Wheatstone's kaleidophone consisted of a circular board about nine
inches in diameter (see figure 4.2). Three perpendicular steel rods were fixed
in the board at equal distances from the circumference and from each other;
each of these was topped with a bauble that was able to reflect light. The first
rod was cylindrical, possessing a diameter of a tenth of an inch. Atop the
rod sat a "spherical bead" (also known as a "steel bead"), a thin glass silvered
on the interior surface, approximately one-sixth of an inch in diameter.[30] A
second rod was topped by a plate that could be adjusted through a range of
angles. The third of the rods emerging from the outer part of the circle was
a four-sided prism, and it too had a plate at its upper end. These plates held

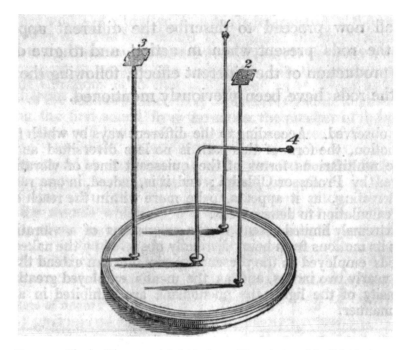

Figure 4.2. Charles Wheatstone's kaleidophone. ("Description of the Kaleidophone, or Phonic Kaleidoscope; a New Philosophical Toy, for the Illustration of Several Interesting and Amusing Acoustical and Optical Phenomena," *Quarterly Journal of Science, Literature, and Art* 23 [1827]: 345.)

colored beads. A fourth rod, attached to the center of the board, was bent at a right angle, and it too had a bead attached at its end.

The kaleidophone was meant to produce both musical pitches and geometrical patterns in the dark when struck by a hammer or bowed. In order to generate figures with the greatest brilliance and distinctness, Wheatstone employed a single light source, such as the sun, a lamp, or a candle. He began his series of experiments by bowing the first rod: the vibrations of light reflected off the bead appeared as circular. He continued by bowing the rod at different places with varying degrees of force, thereby generating different pitches as well as "very complicated and beautiful curvilinear forms."[31] When he exerted pressure on the fixed end of a rod at two opposite points, and when the rod was bowed in the direction of that pressure, the resulting track started out as a line, opened up into an ellipse, and then became a circle. By placing his hand on a portion of the rod below where the bow was applied, Wheatstone noted that the motions decreased rapidly and spiral figures were created.

Figure 4.3. The optical figures produced by Wheatstone's kaleidophone. ("Description of the Kaleidophone," *Quarterly Journal of Science, Literature, and Art* 23 [1827]: 348.)

The kaleidophone experiments (as described by Wheatstone) both satisfied scientific curiosity into the nature of vibrations and rewarded aesthetic attention by arraying pleasing patterns of colored light. Compound figures generated by the vibrating object illuminated by several points of light (such as those from several candles) formed a number of patterns "still more pleasing to the eye."[32] This was the case when he placed a bead on the horizontal plate of the second rod: changing the angle of the plate altered the shapes of the curves (see figure 4.3). He recommended using complementary colored beads to produce the most agreeable effect on the eyes. Causing the third, prismatic rod to vibrate, Wheatstone noted that points of light moved only rectilinearly when the rod was set in motion in the direction of either of its sides. When the motion was applied in an oblique direction, a number of compound curves resulted. The fourth rod, which was bent at a ninety-degree angle, vibrated in two sections, with the vibrations traveling in different directions.[33]

SOUND AND LIGHT

As a device to animate both optical and acoustical properties, Wheatstone's kaleidophone was eminently an instrument of its time and place. Optical illusions and the use of various forms of visual media were becoming increasingly popular in public demonstrations of science throughout the nineteenth century.[34] Indeed, we can understand Wheatstone's acoustical research as continuing the analogy of light and sound put forward by Thomas Young (and discussed by Ellen Lockhart in this volume). In his first published essay, Wheatstone advocated for the polarization of sound (a principle that, incidentally, we now know is incorrect: light waves can be polarized, as Étienne Malus discovered in 1807–1809; sounds waves cannot).[35]

Wheatstone was convinced that his acoustical experiments provided proof of the polarization of sound waves. To demonstrate this theory, he glued a rectangular metal rod to a flat board composed of wood, the rod standing orthogonally to the board. He affixed the bottom portion of a tuning fork to the side of a rod, and rotated it 360 degrees (i.e., in a full circle) to determine which angles produced sound from the board.[36] When the plane of the forks lies along the rod, sound emanates from the board; as the plane of the turning fork turns, the sound decreases until barely audible. At this point the plane of the forks is perpendicular to the rod and therefore parallel to the board's plane. When the tuning fork's plane is parallel to the rod and thus orthogonal to the board's plane, the vibrating fork drives the rod back and forth along its length, thereby making the part of the board to which the end of the rod is glued also move in and out, resulting in the production of sound. When the tuning fork is perpendicular to the rod, the rod vibrates in the same direction. As a result, the attachment point to the board will move back and forth in the plane of the board, producing little or no sound.[37] Wheatstone concluded that in cases where little or no sound is heard:

> The vibrations are thus completely polarized in one direction while passing through the new path, and on meeting with a new right angle they will be transmitted or not, accordingly as the plane of the angle is parallel with, or perpendicular to, the axes of vibrations. In this point of view, the

circumstances attending the phenomena are precisely the same as in the elementary experiment of Malus on the polarization of light.[38]

But were the "circumstances attending the phenomena" of sound polarization really "precisely the same" as that of light? Looking through Iceland spar crystal, Malus had observed that two images illuminated via reflected sunlight from a nearby window would alternately appear and disappear as he rotated the crystal through 360 degrees. Double refraction, which is a consequence of polarization, is the phenomenon by which a single ray of nonpolarized light is split into two rays by an anisotropic medium such as Iceland spar. The two rays travel in different directions: one is refracted at an angle as it moves through the crystal; the other ray moves through unaltered. Wheatstone sought to prove the analogy to sound by means of another simple apparatus of his own devising, again constructed of materials from the music-instrument-builder's workshop:

> When two tuning-forks, sounding different notes by a constant exciter, and making their oscillations perpendicularly to each other, have their vibrations transmitted at the same time through one rod, at the opposite extremity of which two other conductors are attached at right angles, and when each of these conductors is parallel with one of the axes of the oscillations of the forks, on connecting a sounding-board with either conductor, those vibrations only will be transmitted through it which are polarized in the same plane with the angle made by the two rods through which the vibrations pass; either sound may be thus separately heard, or they may both be heard in combination by connecting both the conductors with sounding-boards.[39]

Light thus seemed to provide an explanation for a property of sound. Alas, he produced phenomena in sounding solid "conductors," as he called them, which were merely analogous to the polarization of light. The effect observed by Wheatstone is produced by the transmission of sound through solids, not the polarization of air. However, and although we now know it to be limited, the analogy between sound and light served as evidence for the unity of nature's varied manifestations. The "light figures" of such "philosophical toys," in other words, were doubly beguiling, as much for the way they "proved" the principles of natural unity as for their illusive audiovisual charm.

RESONANCE

Perhaps the most important of Wheatstone's contributions to acoustical theory was his work on resonance, which he defined as the process of causing a body to vibrate by setting a nearby body in motion.[40] According to him, resonance was generated "by means of the undulations which are produced in the air, or in any fluid or solid medium, by the periodical pulses of the original vibrating body—these undulations being capable of putting in motion all bodies whose pulses are coincident with their own, and consequently, with those of the primitive sounding body."[41] Examples of resonance listed by Wheatstone include the vibrations of a string when another tuned to the same frequency is made to vibrate, and the resounding of a drinking glass to the sound of a voice or musical instrument.

In Wheatstone's research on resonance, musical instruments became scientific ones. For example, he noted that when one of the ends of a vibrating tuning fork is brought near a flute's mouthpiece with the apertures closed to produce the same pitch as the vibrating tuning fork, the volume of the fork will be noticeably louder as a result of the resonance of the flute's air column.[42] He then experimented with two flutes placed parallel to each other. One flute was played covering the keys for a C sharp (all lateral apertures were open), while the tube of the second was drawn out so that it could reciprocate tones sounding a semitone lower. This would be equivalent to the flattening of the first flute by the partial closing of the mouthpiece by the lip. He noted that the intensity of the tone could be changed by opening and closing the first hole of the second flute; this effect was caused by the transmission of the waves from the first flute to the second.[43] As is discussed later in this volume, Wheatstone was also interested in foreign musical instruments whose sounds were augmented by resonance. One such was the Javanese *gendèr*, consisting of eleven vibrating metallic plates suspended horizontally by two strings, one passing through each of the two holes in each plate. An upright bamboo tube was placed under each plate, with the length of each bamboo tube ensuring that it would resonate with the lowest frequency of the plate (see figure 4.4).[44]

All of these are examples in which the resonating body sounds at the same pitch as the original body. But Wheatstone also discovered that there are other examples of resonance whereby the resulting sound resonates at a pitch different from the initial sound. He took a tube six inches in length, closed off at one end by a piston, and placed a vibrating tuning fork generating middle C (256 vibrations per second [vps]) at the pipe's open end. By moving the piston inward by three inches, resulting in an air column one half the initial length,

Figure 4.4. Javanese *génder*. ("On the Resonances, or Reciprocated Vibrations of Columns of Air," *Quarterly Journal of Science, Literature, and Art* 3 [1828]: 179.)

he produced a pitch that was an octave higher than the tuning fork (that is, 512 vps). When Wheatstone used a tuning fork with a lower pitch and tubes with small diameters, whose lengths could be shortened to 1/2, 1/3, 1/4, and 1/5 of the initial length, the resulting pitches were the octave, twelfth, double octave, and seventeenth, respectively. Wheatstone concluded, therefore, that "a column of air may vibrate by reciprocation, not only with another body whose vibrations are isochronous with its own, but also *when the number of its own vibrations is any multiple of those of the original sounding body.*"[45]

Resonance occurring at frequencies different from the original sounding body is the basic principle behind the *guimbarde*, or Jew's harp, which is comprised of an elastic steel reed (or "tongue") riveted perpendicularly to a brass or iron frame. This design enables the player to strike the reed relatively easily when the instrument is inserted into the mouth and is supported by the teeth. The numerous possible pitches depend on the dimensions of the mouth cavity as well as the varied motions of the tongue and lips.

Wheatstone continued his acoustical research on the propagation of sound through solid conducting bodies.[46] This work owed much to Jean-Baptiste Biot, with Bouvard and Martin, and to the later work of Chladni on the speed of sound through solids.[47] Wheatstone's experiments were unique, however, since he was studying the transmission of sounds generated by musical instruments. In addition, he demonstrated that sound waves propagating through solid linear conductors over long distances can excite vibrations in adjacent surfaces—ones that are loud enough to be clearly heard. In 1823, François Arago read Wheatstone's paper on this property of resonance to the Académie des Sciences.[48] As ever, the Englishman's experiments on resonance were geared toward popular spectacle, in that they were performed not only in his museum but also at the Royal Institution during the late 1820s. On February 15, 1828, Britain's leading chemist, Michael Faraday, spoke about resonance on Wheatstone's behalf—apparently Wheatstone was rather shy—using the same Javanese *gendèr* discussed in greater detail by

Figure 4.5. Charles Wheatstone's symphonium, ca. 1830.
(Courtesy of the Science Museum, London.)

Davies later in this volume as an example. On March 7, Faraday lectured to an audience at the Royal Institution, once again using material from Wheatstone's research on resonance in air columns; he continued to offer public lectures based on Wheatstone's work during the next two years.[49]

Because of his apprenticeship, Wheatstone had always been fascinated by the sounding boards of various musical instruments, such as pianos, guitars, and violins. The sounding board of an instrument is responsible for enhancing and augmenting the instrument's sounds; its own vibrations force the surrounding particles of air to vibrate, thereby increasing the volume of the sound. As with the enchanted lyre, when a wire is attached to the sounding board of a piano, passed through an insulating tube inserted in a hole in the floor, suspended around a hook in the ceiling of a room one floor below, attached to the lyre, and then fixed at the lower end to the lyre's sounding board, the resulting sounds seem to emanate from the lyre, when in reality they are generated by the piano. In essence, Wheatstone produced a type of musical "circuit" for the propagation of sounds. Two pianos or two harps could even be connected in such a way as to resonate with each other's sounds: two performers in different rooms could play a duet together to two separate audiences, or one could echo the other.[50] In addition, the sounds of one instrument could be simultaneously transmitted to more than one audience. Wheatstone also spoke of using bowed and reed wind instruments for the transmission of sounds.[51] He concluded his piece with the tantalizing suggestion that sound could even be transmitted from one city to another.

> Could any conducting substance be rendered perfectly equal in density and elasticity, so as to allow the undulations to proceed with a uniform velocity without any reflections and interferences, it would be as easy to transmit sounds through such conductors from Aberdeen to London, as it is now to establish a communication from one chamber to another. Whether any substance can be rendered thus homogeneous and uniform remains for future philosophers to determine.[52]

Thus these "circuits" of musical instruments connected by wires would set the stage for Wheatstone's subsequent work on the telegraph later in the 1830s.

VIBRATING REED PIPES AND SPEAKING MACHINES

However, Wheatstone initially felt in the early 1820s that "the transmission [of sound] to distant places, and the multiplication of musical performances,

[were] objects of far less importance than the conveyance of the articulations of speech."[53] The early nineteenth century witnessed the production
of a new class of instrument: wind instruments with free vibrating reeds
attached to pipes. Two examples were the aeolina and aeolodicon, organ-like
instruments with a range of up to six octaves in which the sound was elicited
by a bellows operated by the performer's knee or foot, setting metal reeds in
vibration. They were often used to accompany choirs in regional churches
too poor to afford organs.[54] Aeolinas and aeolodicons were similar to the
more popular physharmonica, a four-octave keyboard instrument with free
reeds invented by Anton Haeckl of Vienna.[55]

Reed pipes were in a sense boundary objects: not only were they important to musical instrument building, but they were also scientific instruments crucial to the study of adiabatic processes in which heat is neither lib-
erated nor absorbed.[56] They were of primary interest to one of the German
territories' leading experimental physicists, Wilhelm Eduard Weber, who
from 1827 to 1830 investigated the physics of vibrating reeds and air columns
with a view to constructing compensated reed organ pipes (ones that retain
the same pitch regardless of volume).[57] In addition, he used such instruments
both to measure the speed of sound in air and to determine the ratio of the
increase in density to the increase in pressure of sound waves. Particularly
relevant to Wheatstone, reed pipes were also critical to the production of
eighteenth- and nineteenth-century speaking machines, which were occasionally attached to automata in order to enhance their appeal.

Wheatstone concluded his essay "On the Transmission of Musical
Sounds" by claiming that the human voice "may be perfectly, though feebly,
transmitted" by connecting a solid object, which would serve as a conductor of sound, to the larynx, or by placing the mouth of a speaker or singer in
close proximity to a sounding board. The key was to communicate sounds
through solid bodies:

> but could articulations similar to those enounced by the human organs of
> speech be produced immediately in solid bodies, their transmission might
> be effected with any required degree of intensity. Some recent investiga-
> tions lead us to hope that we are not far from effecting these desiderata;
> and if all the articulations were once thus obtained, the construction of
> a machine for the arrangement of them into syllables, words, and sen-
> tences, would demand no knowledge beyond that we already possess.[58]

Four years after writing these words, Wheatstone presented his new instrument—a speaking machine—to the British Association for the Advancement

of Science meeting, which took place in Dublin in 1835. As the proceedings reported:

> Professor Wheatstone communicated to the Section an interesting ac-
> count of the various contrivances which have been made to imitate the
> human voice—from the speaking machines of the ancients to those of
> Kempelen and the German mechanists, and the instrument for the pro-
> duction of the vowel sounds contrived by Mr. [Robert] Willis. The Profes-
> sor explained the general principles of this nature, and illustrated their
> effects by experiments. In one of these instruments a pipe, whose length
> could be altered at pleasure by a moveable piece at the end, was made
> to sound by a reed, the air being supplied from a large bellows. By alter-
> ing the length of the pipe while sounding, it was made to give the vowel
> sounds and their various combinations.

A subsequent, improved version of the machine contained "a set of valves, governed by keys, through which the air was admitted to the tube, and partly by the modifications in the form of its mouth which were effected by the hand. The instrument uttered the words 'papa,' 'mamma,' 'summer,' and many others with such distinctness."[59]

He never developed his speech machine further, keeping it at home to entertain dinner guests.[60] He did, however, think that the re-creation of speech artificially by mechanical means (such as with reed pipes) could "show how far the united labours of the philosopher and the mechanician have advanced the inquiry respecting the physical causes upon which these articulations depend."[61] One rather important inventor witnessed Wheatstone's personal demonstration of the speaking machine. At the age of thirteen, Alexander Graham Bell and his father visited Wheatstone. Bell recalled in later life, "I saw Sir Charles manipulate the machine and heard it speak, and although the articulation was disappointingly crude, it made a great impression upon my mind."[62]

REED PIPES, CONCERTINAS, AND SYMPHONIUMS

Reeds were important not only to Wheatstone's speaking machine; they played a critical role in the design of his two inventions, the concertina and the symphonium.[63] These instruments were closely related, the key difference being that the concertina was powered by a bellows while the symphonium was played by the mouth (see figures 4.5 and 4.6). Their precursors undoubtedly were the German mouth organ (or *Mundharmonika*) and the

Figure 4.6. Charles Wheatstone's concertina, 1850s.
(Courtesy of the Science Museum, London.)

English aeolina. Wheatstone's patent of June 19, 1829, was for an improved arrangement of finger keys on his inventions. As he put it:

> In my improved keyed wind instruments the springs are brought so close together that they occupy little more space than in the Aeolina. . . . In fact, eight springs may be placed in the space of an inch and a half, and their corresponding keys may also be brought much closer together than hitherto, and the wind chest made much smaller than has yet been done for a similar number of notes.[64]

Faraday, who was particularly interested in the laws of vibrations of rods and reeds, discussed both the concertina and symphonium in a public lecture entitled "On the Application of a New Principle in the Construction of Musical Instruments," delivered at the Royal Institution on May 21, 1830.[65]

Wheatstone's symphonium was never a success: only around two hundred of them were manufactured. Concertinas did not initially fare much better, with only a hundred sold by 1844.[66] By the 1850s, however, they had become very popular in Britain, their fame owing much to two young virtuosi who had become enamored of Wheatstone's invention. The Swiss-born Guilio Regondi was a child prodigy guitarist who came to London in 1831 and by June 1834 was touring Ireland with the concertina. The *Dublin Evening Post* described his debut, referring to the concertina as "esteemed by fashionable circles of London, the most elegant novelty in the list of musical instruments played upon by ladies."[67] Regondi's tours continued through Britain, the German territories, Vienna, and Prague.[68]

Wheatstone & Co. was the most prestigious manufacturer of concertinas from 1835 to 1870, and its clientele was drawn from the upper classes. Members of the titled aristocracy were well represented on the company ledgers, with aristocratic women outnumbering men by more than two to one even though, on the whole, 88 percent of the buyers of concertinas were men.[69] It is clear that during the 1840s and 1850s, few could afford a Wheatstone concertina. An advertisement from 1848 priced concertinas that could go beyond a beginner's repertory between £14 and £16.16.0, well past what a working-class amateur could afford; even the cheapest concertinas cost £5.15.6.[70] The instrument also fared well with members of the clergy, professional musicians, and particularly instrument dealers and makers.[71] By the 1850s an increasing number of composers began to write for the instrument, with concerts featuring it reviewed more often in the press. Wheatstone changed the tuning from meantone to equal temperament in the 1850s, to match the more general switch in tuning among keyboard instruments.[72]

‹∞›

This chapter has located Charles Wheatstone's early work on acoustics within the context both of London's diverse theaters of wonder, magic lanterns, optical illusions, musical instrument making, and of scientific and technological developments based on experimental study of the properties of vibrations. His demonstrations of acoustical phenomena in his museum displayed all the showmanship typical of the Strand scene at that time. As an instrument maker—a profession with which he closely identified, even after being appointed Professor of Experimental Philosophy at King's College London in 1834[73]—he had a talent for staging exhibitions that showed off both his inventions and scientific acumen to curious Londoners. It is important to note that at the time of his initial academic appointment, the natural and experimental sciences were separate disciplines at King's College. The experimental sciences, which were based on a skilled hand, were well suited to Wheatstone, whose training in the physical sciences came from his family apprenticeship (the natural sciences, by contrast, were seen as a labor of the mind). His is a story similar to that of Faraday, who was from a similar socioeconomic background and who became the first Fullerian Professor of Chemistry at the Royal Institution in 1833. The combination of handiwork and a knowledge of scientific principles evidently assisted those wishing for social mobility. Wheatstone's early lectures were dedicated to the laws of sound, specifically the works of Chladni, Weber, Willis, Faraday, and others.[74] His experiments did much to contribute to the popularization of science, as evinced by his musical museum and a number of Faraday's lectures based on his work. An overview of Wheatstone's career thus situates him in a very real sense at a nodal point in history, before the precarious consolidations of later orders of disciplinary knowledge. Wheatstone worked at the intersection of numerous vibrant cultures, including the worlds of musical instrument making, experimental natural philosophy, speaking machine research, and the public spectacle of science.

NOTES

1. For a biography of Charles Wheatstone, see Brian Bowers, *Sir Charles Wheatstone FRS, 1802-1875* (London: The Institution of Electrical Engineers in association with the Science Museum, 2001), 10.

2. Ibid., 4–7.

3. "Wheatstone's Musical Museum," *Literary Gazette: A Weekly Journal of Literature, Sciences, and the Fine Arts*, June 22, 1822, 396.

4. "Wheatstone's Musical Museum," 396.

5. Bowers, *Sir Charles Wheatstone*, 8; see also Richard D. Altick, *The Shows of London* (Cambridge, MA: Harvard University Press, 1978), 360.

6. "Wheatstone's Musical Museum," 396; see also Altick, *The Shows of London*, 360.

7. "Royal Polytechnic Institution," *Illustrated London News*, February 3, 1855, 117–18, cited in Altick, *The Shows of London*, 387.

8. On situating scientific knowledge, see David N. Livingstone, *Putting Science in Its Place* (Chicago: University of Chicago Press, 2003).

9. *Public Advertiser* (1776), cited in *Notes & Queries*, 11th ser. 3 (1911), 125–26, as quoted in Altick, *The Shows of London*, 63.

10. Altick, *The Shows of London*, 66.

11. Ibid., 69–72.

12. While some have suggested that Merlin opened his museum in 1783, the years 1788 or 1789 seem more likely; see Ann French, "John Joseph Merlin: A Biographical Sketch," in *John Joseph Merlin: The Ingenious Mechanick* (London: Greater London Council, 1985), 14.

13. It is generally thought to have closed its doors five years later; see French, "John Joseph Merlin," 15.

14. Ibid., 16.

15. Simon Schaffer, "Babbage's Dancer and the Impressions of Mechanism," in *Cultural Babbage: Technology, Time and Invention*, ed. Francis Spufford and Jennifer Uglow (London: Faber, 1996), 53–80; available at http://www.imaginaryfutures.net/2007/04/16/babbages-dancer-by-simon-schaffer/ (accessed May 4, 2013).

16. Schaffer, "Babbage's Dancer."

17. Ibid.; see also Tom Standage, *The Turk: The Life and Times of the Famous Eighteenth-century Chess-Playing Machine* (New York: Walker & Co., 2002), 122–28.

18. Simon During, "'The Temple Lives': The Lyceum and Romantic Show Business," in *Romantic Metropolis: The Urban Scene of British Culture, 1780–1840*, ed. James Chandler and Kevin Gilmartin (New York: Cambridge University Press, 2005), 210. See also Iwan Rhys Morus, *Michael Faraday and the Electrical Century* (London: Icon Books, 2004), 17–18.

19. Altick, *The Shows of London*, 66.

20. Mr. Haddock, *A Description of Mr. Haddock's Exhibition of Androides, or Animated Mechanism; Also of the Telegraph, Worked by an Automaton, with Telegraphic Dictionary, etc.* (Dublin: C. Downes, 1800), 3–5.

21. Unidentified newspaper cutting, "Theatre Cuttings: The Lyceum" folder, British Library, cited in During, "'The Temple Lives,'" 221.

22. David Brewster, *Letters on Natural Magic, Addressed to Sir Walter Scott, Bart.* (London: Chatto, 1832), 285–86.

23. For an excellent account of the geographical spaces of nineteenth-century science, see David N. Livingstone and Charles W. J. Withers, editors, *Geographies of Nineteenth-Century Science* (Chicago: University of Chicago Press, 2011).

24. Aileen Fyfe and Bernard Lightman, "Science in the Marketplace: An Introduction," *Science in the Marketplace: Nineteenth-Century Sites and Experiences*, ed. Fyfe and Lightman (Chicago: University of Chicago Press, 2007), 13.

25. See Myles W. Jackson, *Harmonious Triads: Physicists, Musicians, and Instrument Markers in Nineteenth-Century Germany* (Cambridge, MA: MIT Press, 2006), 13–44; and Melissa Dickson, "The Enchanted Lyre," in the present volume.

26. Charles Wheatstone, "New Experiments on Sound," originally published in Thomson's *Annals of Philosophy* 6 (1823): 81–90; reproduced in *The Scientific Papers of Sir Charles Wheatstone, D.C.L., F.R.S.* (London: Taylor and Francis, 1879), 1–13, here, 2–3.

27. Ibid., 5. Wheatstone was told that he was not the first to perform such a correction on Chladni's research. Ørsted had performed similar experiments using alcohol and lycopodion powder (ibid., 7). Wheatstone returned to a more detailed study of the Chladni figures in 1833; see Charles Wheatstone, "On the Figures Obtained by Strewing Sand on Vibrating Surfaces, Commonly Called Acoustic Figures," in *The Scientific Papers of Sir Charles Wheatstone*, 64–83 (originally published in *Philosophical Transactions of the Royal Society* 123 [1833]: 593–634).

28. "Description of the Kaleidophone, or Phonic Kaleidoscope; a New Philosophical Toy, for the Illustration of Several Interesting and Amusing Acoustical and Optical Phe-

nomena," in *The Scientific Papers of Sir Charles Wheatstone* (1879), 21–29 (originally published in *Quarterly Journal of Science, Literature, and Art* 1 [1827]: 344–51).

29. Ibid., 21.

30. Ibid., 23.

31. Ibid., 25.

32. Ibid., 26.

33. Ibid., 27. This research was similar to Antoine Lissajous' subsequent work on optical figures in the 1850s. Unlike Wheatstone, however, Lissajous was able to offer precise measurements of pitches and musical intervals by attaching mirrors to the ends of two resonating tuning forks placed perpendicular to one another and shooting a beam of light, which reflected off the mirrors and onto the screen. The shape of the resulting curves would indicate the musical interval. He used this technique for the creation of the French *diapason normal* at 435 vibrations per second in 1858–1859; see Jackson, *Harmonious Triads*, 210–13.

34. Fyfe and Lightman, "Science in the Marketplace," 7–8, 11–12. See also Lightman, "Lecturing in the Spatial Economy of Science," in *Science in the Marketplace*, 97–113; and Iwan Rhys Morus, "'More the Aspect of Magic Than Anything Natural': The Philosophy of Demonstration," in *Science in the Marketplace*, 336–70.

35. Sound waves cannot be polarized since the direction of the wave propagation and the direction of vibration of the wave's particles are the same; longitudinal waves cannot be polarized. For a history of Malus's work on the polarization of light, see Jed Z. Buchwald, *The Rise of the Wave Theory of Light: Optical Theory and Experiment in the Early Nineteenth Century* (Chicago: University of Chicago Press, 1989), 44–61.

36. While no figure exists to illustrate this, the following paragraph is the most plausible reconstruction of Wheatstone's experiment.

37. Wheatstone, "New Experiments on Sound," 12.

38. Ibid.

39. Ibid.

40. Charles Wheatstone, "On the Resonances, or Reciprocated Vibrations of Columns of Air," in *The Scientific Papers of Sir Charles Wheatstone* (1879), 36–46, here 36 (originally published in *Quarterly Journal of Science* 3 [1828]: 175–83).

41. Ibid.

42. Ibid., 37.

43. Ibid., 38.

44. Ibid., 40.

45. Ibid., 42 (original emphasis).

46. Charles Wheatstone, "On the Transmission of Musical Sounds through Solid Linear Conductors, and on their Subsequent Reciprocation," in *The Scientific Papers of Sir Charles Wheatstone*, 47–63 (originally published in *Journal of the Royal Institution*, 2 [1831]: 223–38).

47. Ibid., 48–49; see also Jackson, *Harmonious Triads*, 37–42.

48. Wheatstone, "On the Transmission of Musical Sounds," 49.

49. Bowers, *Sir Charles Wheatstone*, 21–22.

50. Wheatstone, "On the Transmission of Musical Sounds," 55–57.

51. Ibid., 57–58.

52. Ibid., 62.

53. Ibid.

54. Jackson, *Harmonious Triads*, 99.

55. Ibid., 102.

56. Susan Star and James Griesemer, "Institutional Ecology, 'Translations' and Boundary Objects: Amateurs and Professionals in Berkeley's Museum of Vertebrate Zoology, 1907–39," *Social Studies of Science* 19 (1989): 387–420.

57. Jackson, *Harmonious Triads*, 111–50.

58. Ibid., 62–63.

59. *Proceedings of the Fifth Meeting of the British Association for the Advancement of Science, Held in Dublin, during the Week from the 10th to the 15th^h of August, 1835, Inclusive with an Alphabetical List of the Members Enrolled in Dublin*, 2nd (Dublin: Philip Dixon Hardy, 1835), 96.

60. Bowers, *Sir Charles Wheatstone*, 34–35.

61. Charles Wheatstone, "Reed Organ-pipes, Speaking Machines, etc.," in *The Scientific Papers of Sir Charles Wheatstone* (1879), 348 (originally published in *London and Westminster Review*, October 1837, 14–22).

62. Edwin S. Grosvenor and Morgan Wesson, *Alexander Graham Bell: The Life and Times of the Man Who Invented the Telephone* (New York: Harry Abrams, 1997), 17, quoted in Bowers, *Sir Charles Wheatstone*, 35.

63. Charles Wheatstone, "Wind Musical Instruments, Patent No. 5803 (1829)"; *Concertina Library: Digital Reference Collection for Concertinas*, accessed April 28, 2013, http://www.concertina.com/wheatstone/Wheatstone-Concertina-Patent-No-5803-of-1829.pdf.

64. Charles Wheatstone, "Wind Musical Instruments, Patent No. 5803 (1829)," *Concertina Library: Digital Reference Collection for Concertinas*, accessed April 28, 2013, http://www.concertina.com/wheatstone/Wheatstone-Concertina-Patent-No-5803-of-1829.pdf.

65. Bowers, *Sir Charles Wheatstone*, 36–37.

66. Ibid, 40. On August 7, 1844, Wheatstone obtained another patent for a series of improvements to the construction of the concertina; see Allan W. Atlas, "Ladies in the Wheatstone Ledgers: The Gendered Concertina in Victorian England, 1835–1870," *Royal Musical Association Research Chronicle* 39 (2006): 1–234.

67. Announcement in the *Dublin Evening Post*, June 12, 1834, quoted in Atlas, "Ladies in the Wheatstone Ledgers," 7.

68. Richard Blagrove also contributed to the concertina's fame, creating an impressive repertory for the instrument; Atlas, "Ladies in the Wheatstone Ledgers," 7.

69. Ibid., 10.

70. Ibid., 17.

71. Ibid., 14–15.

72. Ibid., 22.

73. Bowers, *Sir Charles Wheatstone*, 10.

74. Charles Wheatstone, Esq., "Syllabus of a Course of Eight Lectures on Sound," King's College London, Second Term 1834–1835, Ref: K/PP107/5, King's College Archives, London.

CHAPTER FIVE

Charles Wheatstone's Enchanted Lyre and the Spectacle of Sound

MELISSA DICKSON

I have discovered means for transmitting, through rods of much greater lengths and of very inconsiderable thicknesses, the sounds of all musical instruments dependent on the vibrations of solid bodies, and of many descriptions of wind instruments. . . . One of the practical applications of this discovery has been exhibited in London for about two years, under the appellation of "The Enchanted Lyre." So perfect was the illusion in this instance from the intense vibratory state of the reciprocating instrument, and from the interception of sounds of the distant exciting one, that it was universally imagined to be one of the highest efforts of ingenuity in musical mechanism.

—Charles Wheatstone, "New Experiments on Sound" (1823)

In September 1821, a nineteen-year-old Charles Wheatstone attracted public attention with his exhibition of the "enchanted lyre," or acoucryptophone, at his father's musical instrument shop on Pall Mall. In the shop's exhibition room, a small, hand-held replica of an antique lyre with an ornamented keyhole was suspended from the ceiling and, on being ceremoniously wound by Wheatstone, emitted without any apparent human involvement the musical strains of several instruments: a piano, a harp, and a dulcimer. This seemingly supernatural performance lasted for approximately half an hour and was celebrated by the *Weekly Entertainer* as "both brilliant and beautiful," with tones that were "very sweet."[1] The *Literary Gazette* similarly declared that, "however executed," the "music [was] very pleasing, and the effect extraordinary."[2] In reality, this illusory effect was made possible by an emerging understanding of the properties of acoustic waves. The enchanted lyre was, Wheatstone later explained in his scientific papers, a straightforward sounding box, as the brass wires holding it in place passed through the ceiling and were connected to instruments in the room above. When played by unseen (and unheard) musicians, these instruments produced vibrations that traveled down the wires, causing the lyre to "play" as if by magic (see figure 5.1). Far from the aeolian harp of the romantic imagination, which was played spontaneously by the wind and operated as a kind of channel for the voice of nature, this was a scientific demonstration of the capacity of sound waves to travel more efficiently through solid objects than through air. A material, visible conduit for sound waves between musician and auditor, Wheatstone's lyre was, at the same time, a traditional instrument, a musical toy designed for popular entertainment, and a piece of nineteenth-century scientific technology.

The social and cultural meanings of music, as Richard Leppert has argued, have long been shaped not just by hearing but also by seeing music in performance. Audiences hear music, but they also see the bodies and gestures of musicians and the manner in which they interact both with their instruments and with each other. In taking up this tension between music as, on the one hand, a type of organized yet impalpable sound and also, on the other, that which is produced by or from within the human body, Leppert has traced some of the ways in which music's aural and visual presence constitutes both a relation to and a representation of the body:

> Precisely because musical sound is abstract, intangible, and ethereal—lost as soon as it is gained—the visual experience of its production is crucial to both musicians and audience alike for locating and communicating the place of music and musical sound within society and culture. I am suggesting, in other words, that the slippage between the physical activity to produce musical sound and the nature of what is produced creates a semiotic contradiction that is ultimately "resolved" to a significant degree via the agency of human sight.[3]

"Musical sound," in Leppert's formulation, is necessarily "abstract, intangible, and ethereal," and yet principally located in the socially constructed human body.[4] How, then, is such a contradiction resolved when, as in Wheatstone's popular demonstration, the performing body and the labor required to produce the music are removed from both the sight and site of performance (see figure 5.2)? Leppert is right, I think, to call attention to the potential interplay between abstract, invisible sounds and visible, sound-producing bodies. However, that interplay is significantly complicated by the insertion of new scientific knowledge into the performance, which makes the possi-

wire, which I found it convenient to make in the original form
of the experiment, for the sake of portability and facility of
removal; but, if the apparatus be intended as a fixture, it will
be easier and better to employ but one length of wire. The
wire consisted of four portions: the first part touched the
sounding-board of the piano-forte, and reached half-way to
the floor; the second passed through the insulating-tube in
the floor, and terminated when it reached the ceiling of the

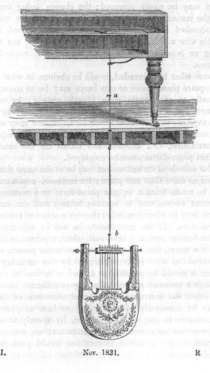

Figure 5.1. Enchanted lyre, or acoucryptophone. (Charles Wheatstone,
"On the Transmission of Musical Sounds through Solid Linear Conductors,"
The Journal of the Royal Institution of Great Britain 2 [Dec. 1831]: 238.
Doe Library, University of California, Berkeley.)

Figure 5.2. Wheatstone's enchanted lyre, or "telegraphic concert."
(James Wylde, *The Magic of Science* [London, 1861], 241. Courtesy
of the California State Library, Sudro Branch, San Francisco.)

bility of experiencing abstract, seemingly magical sounds real. The resulting
formulation of sound as severed—or at least distanced—from its source of
production, and yet newly visible in the form of vibrating strings, renders it
at once abstract and material.

This essay considers the significance of Wheatstone's enchanted lyre; its
context is the emerging science of acoustics in the early decades of the nineteenth century, along with the radical rearrangement of the senses that en-
sued. After an analysis of the newly material nature of sound, in terms of
emerging scientific constructions of its movement through and effects on
the material world, I will turn to the enchanted lyre itself, a device instrumental to the early popular display of those principles. Wheatstone's dem-

onstrations treated sound as a spectacle, interweaving traditions of magic, science, wonder, and showmanship in order to expose and question the capabilities and vulnerabilities of both the human ear and the material object/scientific instrument transmitting and receiving sound. Finally, I will examine the positioning of the lyre as an apparently wondrous conduit for sound waves in motion—occupying a space between source and sound that was the inspiration for myriad scientific claims and cultural fantasies. Charles Wheatstone's enchanted lyre, I argue, made for a striking modern presentation of the nature of sound and the way it travels, positing a new way of listening that became a testament to the wonder of sound as a visual and aural spectacle in the nineteenth century.

THE VISIBLE SOUND

Although it was on one level a simple trick designed to bemuse and entertain, Wheatstone's display of the enchanted lyre was also among a series of early experiments in the transmission of sound; the potential scientific implications of the instrument were far-reaching. This was, the *Literary Gazette* observed, an "affair of sound though not of fury, and signifying something."[5] The inventor himself reportedly declared it to be "entirely the result of a new combination of powers."[6] These exciting conjunctions would make telling contributions to acoustic knowledge, in particular to the theory that sound was propagated by waves or oscillations. Ultimately, Wheatstone conceived of the possibility of transmitting sound across long distances—through his most famous invention, the electric telegraph—by the same means that he had rendered an ordinary lyre "magical."

Fundamental to such aspirations toward new networks of communication and exchange was the need to locate and to physically apprehend sound. The ability to visualize—and thus to measure, record, and quantify—the movement of sound and its effects on objects such as the lyre was a central preoccupation of the burgeoning science of acoustics in the late eighteenth and early nineteenth centuries. In 1787, the German physicist and musician Ernst Florens Friedrich Chladni, in his seminal work *Entdeckungen über die Theorie des Klanges* (Discoveries in the Theory of Sound), had successfully demonstrated that sound impinges on the material world in predictable and scientifically quantifiable ways. Inspired by Georg Christoph Lichtenberg's earlier experiments on electrical figures, in which an electrical spark formed characteristic figures in powder strewn over a non-conducting plate, Chladni hypothesized that a sounding body would vibrate in a similarly systematic

Figure 5.3. Illustration of Chladni patterns in Ernst Chladni's *Die Akustik*
(Leipzig: Brietkopf und Härtel, 1802), n.p.

manner. By drawing a bow over a sheet of metal whose surface was lightly
covered with sand, Chladni revealed that various modes of vibration caused
the sand to concentrate along nodal lines, forming beautiful geometric patterns (now known as Chladni figures) that could be seen with the naked eye
and preserved in drawings (see figure 5.3).[7] In keeping with the later popular reception of Wheatstone's lyre as "enchanted," the Danish physicist and
chemist Hans Christian Ørsted celebrated these figures in sand as objects of
beauty and magic:

Mr. Chladni's experiments are astonishing to anyone who sees them for the first time on account of the regularity of the figures which are produced by a single stroke of the bow, *as if by magic*.[8]

The supernatural charm of Chladni's demonstration—proved by the demonstrator's ability to expose the workings of the natural world—is testament to a new materiality—indeed, a new kind of creativity—invested in the notion of listening. I will return to the dialogue between the magical and the scientific later, but it is important here to note that a new understanding of the movement of sound waves through the world gave the auditory experience a new optical dimension—one with the capacity to generate both wonder and aesthetic pleasure. Chladni's experiments in the substantiation of sound were soon joined by others'. In 1807, for instance, Thomas Young used a stylus to produce direct tracings of the vibrations of sound-producing objects such as tuning forks.[9] Wheatstone's kaleidophone—described in 1827 and named in homage to David Brewster's kaleidoscope—consisted of a series of glass beads and other reflective objects fixed to the ends of rods which, when set in motion by a padded hammer or a violin bow, produced series of spectacular patterns. Like his later wave machine, Wheatstone's kaleidophone helped him to demonstrate the physical presence of sound waves not just in static patterns but in motion.

By way of such devices that created, filtered, represented, and displayed sound, sound itself became a new and exciting part of the material world, implicated not just in the sympathetic vibrations of human bodies but in those of material objects. It moved through space in waves and affected visible change; it had a material trace that could be graphed and analyzed in order to reveal relationships previously unheard and unseen. The visualization of the movements of vibrating bodies made the measurement of frequencies possible: the numeric measurement of pitch was later perfected independently by the German physicist Johann Heinrich Scheibler and the French physicist Jules Antoine Lissajous. In 1833, Wheatstone was able to confirm the existence of sound waves in motion by developing a mathematical formula for Chladni figures. His earlier "Harmonic Diagram," published in 1824, had provided a visual "representation of the principles from which the science of music is derived." It comprised a cardboard disc pinned to a printed card that could be rotated to different settings, indicating or mapping the key signatures of all the major and minor scales.[10] Referring to this diagram as a form of "geographical chart," Wheatstone presented both musical performance and the theoretical understanding of musical sounds as disciplines that might be navigated, recorded, and taught. A material corollary to, or evidence of, an auditory experience became crucial to understanding, visualizing, and communicating it.[11] As Jonathan Sterne has noted, this kind of newly visual sound had a "symbiotic relationship" with the act of quantification:

> Sound had, according to the accepted techniques of science, to be seen in order to be quantified, measured, and recorded; at the same time, some quantified and abstracted notion of sound had to be already in place for its visibility to have any scientific meaning. Again, the product is an artifact of the process: visual sound required the simultaneous construction of sound as a discrete object of knowledge.[12]

In other words, the vibrations of acoustic experimentation, although potentially visible to the naked eye and felt in waves through the body, were only scientifically meaningful when their provenance and the degree of their impact on the material world became known. This knowledge depended on sight.

In this context it is understandable that instruments of both music and science (they were often inseparable in this period) remained the focal point in public demonstrations, enabling scientists to render music visible and audiences to observe it. Press reports on Michael Faraday's lectures to the Royal Institution about Wheatstone's experiments were littered with references to the blades, bows, and tuning forks that were drawn over elastic, metal, or soft wax bodies, and the sands and fluids that were placed on these vibrating surfaces in order to see and chart the results. Each of Faraday's lectures, we are told, was "illustrated throughout by the performance of all the experiments referred to."[13] Objects of scientific investigation and musical performance thus became points of intersection between research, public oration and display, and an early culture of consumerism. Although human performers were still required both to operate these instruments and to interpret the results, the production of sound in general—and musical sounds in particular—could be isolated as observable scientific facts, seemingly removed from the workings of the body and (re)located within the material realm of vibrating and moving things. The scientist/inventor was, in this way, configured as a kind of showman in his presentation of acoustical phenomena: by direct appeal to the senses through a series of visual and aural spectacles and sensations.

Musical instruments became fundamental to this new mode of perceiving sound: many of Wheatstone's experiments explicitly related to the vibrating sounding boards of string instruments and the transmission of columns of air in wind instruments. Far from an intangible, abstract medium, then, the music of the lecture hall and the scientific experiment was experienced in terms of its newly traceable movements in the material world. During one lecture, for example, Wheatstone's accordion, his symphonium, a harp, and an aeolian organ were demonstrated in performance as various applications of a new principle of springs, or tongues, in the construction of musical instruments.[14] At another, "very beautiful Aeolian tones were produced from an instrument which Mr. Faraday had" in order to demonstrate the sound effects produced by air blown through its apertures.[15] Other experiments that did not use specifically "musical" instruments—such as an experiment demonstrating the duration of luminous impressions, which involved a flame of hydrogen gas burning in a glass tube—might still be noted to produce "musical sound."[16] While it is not clear what exactly distinguished musical from nonmusical sound, such demonstrations reinforced and then spectacularized the troubled aesthetic relationship between the mechanical production and scientific measurement of sound waves on one hand, and the "beautiful Aeolian tones" provoking affective response among the audience on the other. It seems that the scientific abstraction of musical activity situated "sound" firmly within the workings of the material world and illuminated its inherent need for ongoing experimentation and demonstration in order to create, re-create, and, eventually, record this sound. Similarly, the music produced by Wheatstone's enchanted lyre was "calculated to afford the highest gratification to the Musical World"; the *Literary Gazette* praised the "beauty of the effects produced" while recognizing that it was enmeshed in theatrical and mechanical display.[17] This emerging tension between the abstract and the material provided significant scope, I would suggest, for individual and cultural fantasies of enchantment and supernatural presence. I want to turn now more fully to Wheatstone's lyre in order to illustrate the ways in which the instrument afforded popular opportunities to engage with sound as a source of wonder and spectacle.

POPULAR SOUND

In an 1821 review of Wheatstone's demonstration of the enchanted lyre, *Ackermann's Repository* remarked on the striking audiovisual effect of this supposedly self-performing instrument. The writer emphasized the crucial sense of secrecy that underpinned the spectacle:

> It is evident that some acoustical illusion, effected through a secret channel of some sort or other, is the cause of our hearing the sound in the

belly of the lyre. The lyre augments no doubt the vibration, but in other respects it seems to act as a mere representative: any other vibrating receptacle of a different shape would probably answer the inventor's purpose equally well. How, then, is the sound thus conducted so as to deceive completely our sense of hearing? This seems to be the only question that can suggest itself on witnessing this singular experiment; it is the secret upon which Mr. Wheatstone rests the interest and merit of this invention; and to this question, no one, as far as we could learn, has yet been able to return that answer that could solve every difficulty.[18]

The vibrating strings of the suspended lyre render its music a palpable and visible phenomenon as well as an audible one. Through vibrations, the audience was induced to see, hear, and potentially feel the sounds being transmitted and speculate as to the origin of the transmission and the nature and importance of the receiving vessel. The lyre became a physical object of contemplation, one that substantialized sound as a medium of heavenly provenance. It could be approached and analyzed by the mind and by the senses, but it also concealed a mystery that could not easily be explained by the viewer.

The *Repository* reviewer's anxious notion that there might be an element of deception or trickery in this process is, I think, significant, as it implies a deepening awareness of the existence of sounds and movement—natural, human, or supernatural—beyond the limits of ordinary sensory perception. This was an approach shared by the *New Monthly Magazine*:

> We are convinced at once that it is not the lyre that gives us the musical treat, but that a skillful player is somewhere else occupied in entertaining and puzzling us. Nevertheless, on approaching the lyre, and holding the ear close to it, we are equally assured that the sound proceeds from the belly of the lyre itself. In this dilemma we are left to conclude that the sound is conducted into the lyre; but the means of this harmonic introduction have as yet eluded the most minute investigation.[19]

It was an evident source of confusion for these reviewers that the network of senders and wires, recipients and relays composing the exhibition of the enchanted lyre could not fully be comprehended through sight or sound alone. The implication of such an experience was that there was more to be heard, felt, and seen beyond everyday sensory thresholds. Nearly forty years later, in a public lecture delivered in 1857, the German physician and physicist Hermann von Helmholtz described the limitations of the human ear in perceiving the musical scale, identifying the point in the lower register at which the listener became aware of multiple vibrations rather than notes as "the ear refuses its office, and hears slower impulses separately, without gathering them up into single tones."[20] While reviews of the lyre exhibited less technical knowledge of the science of acoustics, they nevertheless betrayed consciousness of the potential for discovering such further realities, beyond audible sound. These new realities were, of course, quite literally available for discovery in the case of Wheatstone's exhibition: the notes propagated along the brass wires caused sympathetic vibrations of the lyre's strings while the sounds of the piano filling the air in the room above did not reach the audience's ears, thus demonstrating that sound waves travel more efficiently through solid objects than through air. As the appellation "acou-cryptophone" suggests, this was a visual and aural spectacle that separated source from sound, the unheard from the heard, the extrasensory from the sensory—a testament to the wonder and the sheer novelty of new scientific principles at work.[21]

Drawing his investigations firmly into the realm of London's "shows" culture, Wheatstone's highly theatrical mode of presentation involved the use of an ornamental keyhole with a clock mechanism on the front of the lyre, a useless item that he would occasionally wind up at the start of the performance. The lyre's subsequent one-hour concert was, in this context, a musical amusement "in the most superior style" and also a piece of theater.[22] The dynamic cofunctioning of science and showmanship thus involved a type of "modern enchantment" (to use Simon During's helpful expression). Magic, During argues, had "slowly [become] disconnected from supernature" to persist in modern culture as the "self-consciously illusory" performance of conjuring acts, new technologies, stage productions, and special effects.[23] The enchanted lyre, I would suggest, facilitated an early example of such "modern enchantment." Here, magic and science, reason and irrationality were imbricated in ways that were not alien but rather central to British scientific and technological modernity. Even though no serious claim to the supernatural was being made, the performance presented a delicate balance between science and mystery, a state that imbued the very act of listening with a magical effect.

Wheatstone's enchanted lyre might usefully be compared with the same inventor's "Apparatus of the Invisible Girl," which was exhibited around the same time and again imbued the transmission of sound—in this case, the human voice—with quasi-magical properties. A relatively simple contrivance which had appeared periodically in various guises in Britain and America during the first decades of the nineteenth century, the exhibition comprised a wooden frame holding four trumpets into which, on the showman's invitation, a spectator would pose a question and then receive an answer from the invisible girl, an oracle supposedly contained within the body of the trumpet itself. In reality, as Richard Altick explains, a natural rubber tube was threaded through the hollow frame, conveying sounds to a woman hidden in an adjoining room. A concealed hole in the partitioning wall enabled her to see and describe the spectators, supposedly convincing them of her intimate knowledge of their lives.[24] Like the enchanted lyre, the invisible girl's commercial success depended on hints of a vaguely supernatural presence hidden within the mechanical/scientific process, one detached from human existence. In this case, the four trumpets became conduits between the known and unknown, portals into enchantment. This experience, too, made no claim to the supernatural and was always known to have been manufactured; but it nevertheless produced effects of supernatural immediacy that allowed the viewer to forget the presence of the medium. In Irish poet Thomas Moore's sentimental tribute "To the Invisible Girl," first published in 1803, the speaker moves from the seen, material trumpet to the unseen, supposedly ethereal presence it contains, celebrating the girl as an emblem of "consoling enchantment" and evoking a tension between the "sweet spirit of mystery" and the epistemological certainty provided by scientific analysis. Despite its lighthearted tone, the poem's boundaries between magical phenomena and scientific possibilities are disturbingly blurred in a nostalgic reflection on the joys of simple fancy:

> They try to persuade me, my dear little sprite,
> That you're *not* a true daughter of ether and light,
> Nor have any concern with those fanciful forms
> That dance upon rainbows and ride upon storms;
> That, in short, you're a woman; your lip and your eye
> As mortal as ever drew gods from the sky.
> But I *will* not believe them—no, Science, to you
> I have long bid a last and careless adieu:
> Still flying from Nature to study her laws,
> And dulling delight by exploring its cause,
> You forget how superior, for mortals below,
> Is the fiction they dream to the truth that they know.[25]

Moore both knows and regrets that, unlike the sprites of his fancy, this unseen woman possesses a lip to communicate with her audience and an eye to observe without being observed; his resistance to the supposed truth of Science manifests itself as a self-conscious rejection of the corporeal in favor of the imagination, magic, and the realms of fairies and spirits. In other words, there is here a complex interaction between text, transmission, object, and technology, as the wonder of a magic that reveals and manipulates the material world becomes the wonder of science itself. The repeat performances of the "Apparatus of the Invisible Girl," like those of the enchanted lyre, not only bear witness to the wonder, novelty, and imaginative possibilities of scientific principles at work; their success also betrays the extent to which magic and science were interwoven at the foundation of British modernity.

Such a magic of the mechanical has recently been taken up by Jeffrey Sconce in the context of electrical experimentation. Sconce traces the metaphors of spiritual "presence" that electrical technologies have spawned, from the emergence of the nineteenth century's electric telegraph to the virtual realities of the twenty-first century, and claims that a persistent, causal relationship exists between electricity and the paranormal. Communication technologies, he implies, generate a powerful sense of removal from the material world and give rise to experiences attuned to realms beyond normal consciousness and corporeality. Sconce points out that objects of communications technology are often imagined as magical in the popular consciousness, apparently because of the otherworldly realms they animate.[26] It is indeed tempting to envisage an enchanted lyre that plays by itself or an invisible female presence that holds a conversation through the bell of a trumpet as antecedents to the occult traditions of séances, spirit circles, automatic writing, telepathy, and clairvoyance later in the century. However, I would argue that these apparatuses not only served as an interface between constructions of presence and absence, and between material and immaterial realms, but they provoked speculations upon the nature of that interface, and the potential for movement through it. The newly visible, tangible presence of sound was a source of wonder; it raised profound questions about the capabilities of the human body, as well as about the matter of transmitting and receiving sound in the modern age. As such, the magical effects of Wheatstone's enchanted lyre and invisible girl emphasized the new powers and the restrictions of both acoustical science and the human sensorium: the possibilities of hearing, seeing, and feeling beyond "normal" sensory thresholds.

IMAGINARY SOUNDS

The enchanted lyre, like the invisible girl, provided a novel model of communication, one that extended beyond the limitations of the human senses, straining toward that "roar on the other side of silence" later posited by the

narrator of George Eliot's *Middlemarch* (1871–1872).[27] Scientific speculation
on future uses of traveling sound was rife from the very beginning. Wheatstone himself noted that his self-playing lyre operated according to general
principles for propagating sound through space and solids; he looked forward
to conducting the sounds of wind, rather than string instruments, in similar performances in the future.[28] In his 1831 paper "On the Transmission of
Musical Sounds through Solid Linear Conductors, and on their Subsequent
Reciprocation," Wheatstone published the results of many experiments (including that with the lyre), and discussed the limitations of the transmission
of sound waves over long distances. He made no mention of electricity, but
observed:

> Could any conducting substance be rendered perfectly equal in density
> and elasticity so as to allow the undulations to proceed with a uniform
> velocity without any reflections and interferences, it would be as easy to
> transmit sounds through such conductors from Aberdeen to London as
> it is now to establish a communication from one chamber to another.[29]

This scientific fantasy of sound waves that might be sustained and carried
across vast geographical spaces was shared by *Ackermann's Repository*,
which excitedly envisaged a time when an opera performed at the King's
Theatre might, by way of vibrations traveling through underground cables,
be enjoyed across not only the physical and geographical but also the social
spaces of the metropolis. In this imagining, the physical labors of the emerging figure of the orchestral conductor would be supplemented by physical
vibrations that would conduct music across greater London, creating a new
kind of immediacy in the reception of musical and political events, and thus
producing a unifying national effect:

> Who knows but by this means the music of an opera performed at the
> King's Theatre may ere long be simultaneously enjoyed at the Hanover-
> square Rooms, the City of London Tavern and even at the Horns Tavern
> in Kennington, the sound travelling, like the gas, through snug conduc-
> tors from the main laboratory of harmony in Haymarket to distant parts
> of the metropolis. . . . And if music be capable of being thus conducted,
> perhaps the words of speech may be susceptible of the same means of
> propagation. The eloquence of counsel, the debates of Parliament, instead
> of being read the next day only,—But we shall lose ourselves in the pur-
> suit of this curious subject.[30]

The idea that pursuing this subject might prompt the writer and his readers to "lose ourselves" emerges (at least in part) from the fear that such "means of propagation" would render the human subjects invisible to one another and, at the same time, make human bodies strangely amenable to heightened and altered sensory states. There is a paradoxical coupling of immediacy with detachment here: new technologies allow sound to be communicated across vast distances in an instant, connecting societies and individuals; but in the process they emphasize the gap separating transmitting and receiving bodies.

Cultural fantasies of communication, at great distance or indeed beyond sensory thresholds, frequently drew on both science and magic in order to broker the boundaries of the individual self and its range of sensory perception. Shelley Trower has recently taken up the notion of the vibration as a material experience of sound, casting nineteenth-century spiritualism as a "dream of capturing vibrations, of engineering a kind of controlled access to the realms of extrasensory frequencies."[31] In this context she identifies a little-known poem by hymn writer Frances Ridley Havergal in which the romantic image of the aeolian harp operates as a device for detecting extrasensory vibrations. In "The Message of the Aeolian Harp," written in 1869, the widowed Eleanor believes that the "music of [her husband's] life" continues even though "our poor ears no longer hear it."[32] Her harp operates like the trumpets through which the invisible girl speaks, receiving and transmitting messages from the dead that are unavailable to the human ear. Havergal's poem described a low note "trembl[ing] out of silence":

> It seemed to die; but who could say
> Whether or when it passed the border-line
> 'Twixt sound and silence? for no ear so fine
> That it can trace the subtle shades away;
> Like prism-rays prolonged beyond our ken,
> Like memories that fade, we know not how or when.[33]

Trower reads this hymn in order to focus on the nature of sound as "a kind of energy that fades away beyond one's limited powers of sensitivity, but may continue to exist," in the process offering a compelling reading of the harp's vibrations as consoling spiritual vibrations that transcend the confines of space and time.[34] I would add, however, that the harp itself plays a pivotal role by figuring that liminal, transcendent space between "sound and silence." Like Wheatstone's lyre, it is a tangible conduit in an invisible system of transmission and reception; it generates wonder both for the persistent strains of magic in Western culture and for the technological and scientific advancements of British modernity.

Similarly wonderful objects that capture and transmit sound were conceived throughout the nineteenth century and beyond. For example, in Arthur Conan Doyle's "The Japanned Box" (1899), a phonograph carries the clear, crisp voice of a deceased woman; or in Rudyard Kipling's "Wireless" (1902), mechanical signals inadvertently channel the creative spirit and poetry of the long-dead poet Keats; or in Florence McLandburgh's "The Automaton Ear" (1873), an unnamed professor invents a device that enables him to detect sounds beyond the limits of the human ear, only to be haunted by the cries of the dead. In each case, the scientific instrument in question establishes a threshold between life and death, the physical and metaphysical, offering simultaneously exhilarating and devastating possibilities of movement between the two. The human body or, more precisely, the human ear, becomes peculiarly vulnerable to sound beyond that threshold. In the opening pages of McLandburgh's tale, the nameless professor comes across the following lines (which bear a remarkable resemblance to Charles Babbage's reflections in his 1838 chapter "On the Permanent Impression of Our Words and Actions on the Globe We Inhabit" [35]) while reading under a tree:

> As a particle of the atmosphere is never lost, so sound is never lost. A strain of music or a simple tone will vibrate in the air forever and ever, decreasing according to a fixed ratio. The diffusion of the agitation extends in all directions, like the waves in a pool, but the ear is unable to detect it beyond a certain point. It is well known that some individuals can distinguish sounds which to others under precisely similar circumstances are wholly lost. Thus the fault is not in the sound itself, but in our organ of hearing, and a tone once in existence is always in existence.[36]

Aspirations for transcendence and a heightened extrasensory experience drive the scientist's inventions. The professor's desire for immediacy draws the supernatural at least partially into the realm of human sensory perception by way of those scientific instruments that record and filter sound, while emphasizing the fragility and mortality of the human body.

In another recent cultural history, Richard Menke reads the Victorian realist novel as profoundly shaped by its interactions with the penny post, telegraph, and wireless receiver—all systems of communication that brought distanced individuals into contact with one another by way of increasingly rapid communication across material networks. Charlotte Brontë's *Jane Eyre*, for example, is for Menke a novel "charged with the spirit of electric telegraphy, a spirit that accounts from the late 1840s associate with the sympathy and synchronicity between separate lovers"; the novel posits a fantasy of telecommunication that surpasses the capacities of actual information systems in 1847.[37] Interestingly, however, rather than using a scientific instrument to engineer this fantasy, Brontë treats the physical bodies of Jane and Rochester as themselves both material conduits and supernatural entities in the network of exchange. Confronted with cousin St. John Rivers's demand that she accept his proposal of a loveless marriage and accompany him as a missionary to India, Jane hopes for a sign of divine intervention. In response, she hears the voice not of God but of her estranged and far-distant lover, Mr. Rochester. This preternatural, disembodied sound establishes a deep psychological connection between Jane and her beloved, one later confirmed by Rochester as an accurate transmission of his cry for her presence:

> "What have you heard? What do you see?" asked St. John. I saw nothing: but I heard a voice somewhere cry: "Jane! Jane! Jane!" Nothing more. "O God. What is it?" I gasped. I might have said, "Where is it?" for it did not seem in the room—nor in the house—nor in the garden: it did not come out of the air—nor from under the earth—nor from overhead. I had heard it—where, or whence, for ever impossible to know! And it was the voice of a human being—a known, loved, well-remembered voice—that of Edward Fairfax Rochester; and it spoke in pain and woe.[38]

In 1847, then, Brontë anticipated both telephony and the turn-of-the-century séance in order to, in Menke's words, "imagine an experience of silent speech across the miles"—a similar fantasy, one might add, to Wheatstone's dreams of immediate long-distance communication.[39] Building on Menke's suggestion of the technological foresight contained here, I would emphasize the fundamental urge for physical contact running through this strange moment of exchange between characters. Significantly, this passage represents a kind of climax to a sequence of episodes involving powerful auditory stimulation in the novel. As a girl on the verge of puberty, locked in the red room by her aunt, Jane is fearful that her "violent grief might awake a preternatural voice to comfort me,"[40] and while this voice does not make itself heard, after her terrified fit she claims that a voice "came out of me over which I had no control."[41] Later, as governess at Thornfield, she is possessed by her own inner voice, that "secret voice which talks to us in our own hearts," and also regularly "haunted" by a manic, disembodied laugh that she struggles to assign to the physical presence of Grace Poole. Each of these auditory experiences occurs during moments of physical or mental isolation; there are clear emotional, sexual, and psychological dimensions implicit in Jane's responses to these stimuli. These situations of inner hearing paradoxically reinforce the importance of physical human contact and the obdurate materiality of the (in this case, absent) body. The supernatural/electrically charged exchange between Jane and Rochester thus draws on contemporary telegraph discourses while expressing a deep desire to obviate the need for telegraphy: by realizing the actual physical and intellectual intimacy that technology simulates. Again, the limitations and capabilities of the body to hear sound are brought to the fore, as the psychic connection and immediacy of disembodied exchange emphasizes the separation and the physical yearning of bodies in a telegraphic circuit. Mechanical, supernatural, and bodily functions evoke, reflect, and comment on one another, exposing the recalcitrant materiality of nineteenth-century fantasies of disembodiment.

Charles Wheatstone's 1821 exhibition of a lyre, held not by a hand but by a brass wire, immediately draws our attention to the potential replacement or the potential enhancement of human contact by way of mechanical, scientific, or magical means. By removing his performers from the site of sound production, a philosophical space was pried open between the seen and unseen, the heard and unheard, in which a new aesthetics of wonder and creative possibilities emerged. In the absence of a visible musician or speaker, moreover, previously unheard worlds of auditory experience were made available for acoustic study, wherein sound was substantialized as a matter of concern; that is, the abstracted physicality of sound became a provocation for new enchantments and scientific speculation. An object at once of spectacle and of science, the enchanted lyre itself became both a tangible model of sound waves in action and a physical demonstration of the limitations of human sensory perception.

NOTES

1. "Varieties," *Weekly Entertainer and West of England Miscellany* 14 (1821): 222.

2. "Arts and Science," *Literary Gazette* 243 (1821): 586.

3. Richard Leppert, *The Sight of Sound: Music, Representation, and the History of the Body* (Berkeley: University of California Press, 1993), xx–xxi.

4. Ibid., xx.

5. "Sketches of Society," *Literary Gazette* 270 (1822): 185.

6. "Arts and Sciences," *Literary Gazette* 243 (1821): 586.

7. For a more detailed outline of Chladni's research into acoustics, see Jonathan Sterne, *The Audible Past: Cultural Origins of Sound Reproduction* (Durham, NC: Duke University Press, 2003), 43–45.

8. Hans Christian Ørsted, "A Letter from Mr Ørsted, Professor of Philosophy in Copen-

hagen, to Professor Pictet on Acoustic Vibrations [1805]," in *Selected Scientific Works of Hans Christian Ørsted*, ed. and trans. Karen Jelved, Andrew D. Jackson, and Ole Knudsen (Princeton, NJ: Princeton University Press, 1998), 182 (emphasis added).

9. See Ellen Lockhart's essay in the present volume.

10. "Music," *La Belle Assemblée: Or, Court and Fashionable Magazine*, May 1824, 221.

11. Lisa Gitelman has made a similar claim regarding the American public's later reception of Edison's phonograph, which was designed to capture the sounds of speech in a new recorded form by way of a stylus that traced sound vibrations on a rectangular sheet of tinfoil wrapped around a rotating cylinder. These fragile, indented sheets, which were entirely useless without the phonograph to provide playback, were regularly taken home by members of the audience as souvenirs of their auditory experience and, Gitelman contends, of a new technology that made visible certain anxieties regarding the medium of print. See Gitelman, "Souvenir Foils: On the Status of Print at the Origin of Recorded Sound," in *New Media, 1790-1915*, ed. Lisa Gitelman and Geoffrey B. Pingree (Cambridge, MA: MIT Press, 2003), 157-73.

12. Sterne, *The Audible Past*, 44.

13. William Jerden, ed., "Arts and Sciences: Royal Institution; Musical Sounds," *Literary Gazette* 592 (May 1828): 330.

14. William Jerden, ed., "Arts and Sciences: Royal Institution," *Literary Gazette* 698 (June 1830): 369-70.

15. Jerden, "Arts and Sciences: Royal Institution; Musical Sounds," 330.

16. "Royal Institution," *Athenaeum* 281 (March 16, 1833): 171.

17. "Wheatstone's Musical Museum," *Literary Gazette* 283 (June 1822): 396.

18. Rudolph Ackermann and Frederic Shoberl, "Musical Intelligence: The Enchanted Lyre," *Ackermann's Repository of Arts, Literature, Fashions, Manufactures, &c.*, 2nd ser., vol. 12 (London: R. Ackermann, Strand, September 1, 1821): 174.

19. "Music," *New Monthly Magazine and Literature Journal* 17 (1822): 200-1.

20. Hermann von Helmholtz, "On the Physiological Causes of Harmony in Music," *Science and Culture: Popular and Philosophical Essays*, ed. David Cahan (Chicago: University of Chicago Press, 1995), 48.

21. Wheatstone enjoyed concocting new words. Derived from the Greek, "acoucryptophone" literally means "hearing a hidden sound." See Brian Bowers, *Sir Charles Wheatstone FRS, 1802-1875* (London: The Institution of Electrical Engineers in association with the Science Museum, 2001), 8.

22. "Wheatstone's Musical Museum," *Literary Gazette* 283 (June 1822): 396.

23. Simon During, *Modern Enchantments: The Cultural Power of Secular Magic* (Cambridge, MA: Harvard University Press, 2002), 14-27.

24. Richard Altick, *The Shows of London* (Cambridge, MA: Harvard University Press, 1986), 353.

25. Thomas Moore, *Poetical Works*, 10 vols. (London: Longman, Orme, Brown, Green, and Longmans, 1840-1), 1:33-34.

26. Jeffrey Sconce, *Haunted Media: Electronic Presence from Telegraphy to Television* (Durham, NC: Duke University Press, 2000).

27. George Eliot, *Middlemarch*, ed. Rosemary Ashton (Harmondsworth, UK: Penguin, 2003), 194.

28. Ackermann and Shoberl, "Musical Intelligence," 174.

29. Charles Wheatstone, "On the Transmission of Musical Sounds through Solid Linear Conductors, and on their Subsequent Reciprocation," *The Scientific Papers of Sir Charles Wheatstone*, 62.

30. Ackermann and Shoberl, "Musical Intelligence," 174.

31. Shelley Trower, *Senses of Vibration: A History of the Pleasure and Pain of Sound* (New York: Continuum, 2012), 72.

32. Cited in ibid., 72.

33. Cited in ibid., 72.

34. Ibid., 72.

35. Charles Babbage, "On the Permanent Impression of Our Words and Actions on the Globe We Inhabit," *The Ninth Bridgewater Treatise: A Fragment* (London: John Murray, 1838), 108–19. James Emmott notes this similarity between Babbage and McLandburgh in his "Parameters of Vibration, Technologies of Capture, and the Layering of Voices and Faces in the Nineteenth Century," *Victorian Studies* 53, no. 3 (2011): 468–78.

36. Florence McLandburgh, "The Automaton-Ear," *Scribner's Monthly* 5, no. 6 (April 1873), 711.

37. Richard Menke, *Telegraphic Realism: Victorian Fiction and Other Information Systems* (Stanford, CA: Stanford University Press, 2008), 87.

38. Charlotte Brontë, *Jane Eyre*, ed. Richard J. Dunn, 3rd ed. (New York: Norton, 2001), 419–20.

39. Menke, *Telegraphic Realism*, 82.

40. Brontë, *Jane Eyre*, 17.

41. Ibid., 19.

CHAPTER SIX

Instruments of Empire

JAMES Q. DAVIES

CONTACT INSTRUMENTS

In this chapter, my proposal is to handle musical instruments as a class of communication technology, or rather the other way around, communication technologies as a class of musical instrument.[1] The kinds of instrument that I am interested in are pictured in figures 6.1–6.4. These objects are what I call "contact instruments": typewriter-cum-pianos, telegraph-cum-pianos, concertina-cum-rotary dial telephones. My provocation here is to take an organological view of "modern" telecommunication systems. It is to describe the instruments, both musical and scientific, made possible by and conceived for the purposes of British colonial encounter and conquest. The aim is to show how nineteenth-century instruments of the sort pictured here played into the space of empire, how in fact they *made* that space by shaping knowledge of it. I am interested in how technologies act in engineering physical landscape, how men engaged in the active placement of land through the active use of instruments, and the worlds that Londoners built.

In the previous chapter, Melissa Dickson took a close-up view of Charles Wheatstone's enchanted lyre, which was displayed under the fashionable Royal Opera Arcade on Pall Mall. As Dickson and Jackson both describe in this book, Wheatstone's telephonic precursors to his telegraphic work propagated sound through wires from one soundproofed room to another. Dickson quoted an 1821 reviewer extolling the acoucryptophone's potential to collapse distance, the first of many to imagine a utopian future where "a song sung at the Opera House might be heard at all the other theaters in London."[2] Figure 6.1 shows that Wheatstone also built typewriters for speeding communication, skeuomorphs that looked suspiciously like pianos; other London musicians built such successful apparatuses as the printing telegraph (figure

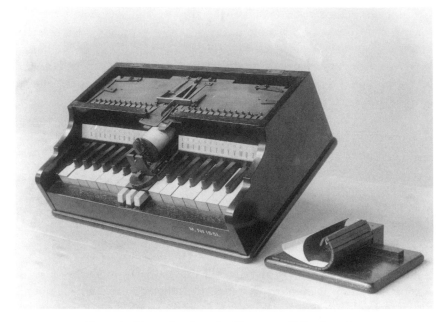

Figure 6.1. Typewriter from the Wheatstone Collection, 1851.
(Science Museum, London/Science & Society Picture Library.)

6.2), the keyboard providing a familiar interface for well-practiced pianists.
Two more Wheatstone communication machines appear in figures 6.3 and
6.4: the first, a telegraph system anticipating the dial telephone, the rotary
design of which, apparently, resembled a laboratory apparatus that Wheatstone, in 1835, was calling "the concertina."

Figures 6.5 and 6.6 involve much larger assemblages not, I think, unrelated
to Wheatstone's acoucryptophone and the effort to enchant sound. These immense objects were conceived by the chief engineer of the Great Western
Railway, Isambard Brunel, who in 1839 collaborated with Wheatstone in the
laying of the *first permanent line of electric telegraph* between Paddington
and Drayton stations in London. Brunel would later turn his attention to the
rising sun by engineering the *Great Eastern*. The *Great Eastern* was a gigantic, eighteen-thousand-ton steam-sail-paddle ship, by far the largest yet constructed, which served the Eastern Telegraph Company by laying out cables
for a Wheatstone submarine telegraph that connected Europe to the Ameri-
cas in 1865, London via Egypt to Bombay in 1866, and then beyond (figure 6.6).
I am thinking what it would mean to imagine Brunel's *Great Eastern* laying
out cables for a vast inter-continental undersea musical instrument.[3]

I have selected these images in order to clarify certain continuities between scientific and musical instruments. Are these instruments for entertainment? Are they for scientific research? Are these technologies instruments of violence, conceived for colonial expansion? Are they useful for what Victorian imperialists might have triumphantly called "the annihilation of distance"?[4] Another reason to incorporate them is to stress the extent to which the history of musical instrument manufacture is enmeshed in the ravaged landscapes that envelop us. We are, in a sense, surrounded by musical instruments, if we properly account for the history of communication technologies. Extending themselves across the seas, Britons imagined those wired environments bequeathed to us by the nineteenth century—beneath, above, and around us—as if a vast imperial "nervous system." Still today, our networked worlds bear witness to the strange conviction that it is only by wiring local space that we apparently get to tune into a pure global beyond.

In sum, my claim is that musical instruments—"contact instruments"—

Figure 6.2. "Printing telegraph," 1855. This proto-telex machine, or teleprinter, was devised by David Edward Hughes, who was born to a family of London musicians à la Wheatstone, a child protégé-sensation on the Wheatstone concertina, professor of music, and inventor of the carbon microphone. (Science Museum, London/Science & Society Picture Library.)

Figure 6.3. Prototype telegraph transmitter (ca. 1850) by Charles Wheatstone,
sporting thirty ivory concertina-style keys on a concertina-style octagonal lid.
(Courtesy of Frances Pattman, King's College London.)

act in the configuration of political geographies. They are levers wherewith
to move the world. They have been deployed in order to shape shared emotional and political space. These instruments, so their advocates say, fold
oceans into continents and configure near and far. As a corollary to the aesthetic sense proper to them, instruments prove concepts, which is to say that
basic musical principles become evident in the playing of them. Particularly
when they become so "normal" that we stop noticing them, standard instruments become productive of standard vocabularies and standard truths.
Like scientific instruments, musical instruments carry with them their own
sense of objectivity, though that sense can never be extrapolated from the instrument alone, since objectivity only emerges by force of use. These instruments, in other words, are implicated in the literal production of land; they
make land by acting in the crystallization of both a politics and a cosmos. We
could speak of a cosmos of standard objects, or what Isabelle Stengers would
call a cosmopolitics.[5]

In the remaining pages of this chapter, I chronicle the search, under
British imperialism, for an Instrument of Instruments, that is, an imperial
standard for music making. I am interested in the quest for this instrument—

a machine capable of "speaking" a universal musical language—conceived for the purposes of annihilating distance in the fashion of the violently imperial projection of global space (see figure 6.6). The earth-moving Ur-Instrument would be founded on "the True Scale," or "The Scale of Nature." It would be amenable to the performance of all known global scale systems. And it was pursued with zeal, as we shall see, by a coterie of popularizing nonconformist scientists, music theorists, reform-minded London evangelicals, and missionaries. This impossible object would be a purely theoretical implement of such power and reach, and such invisibility, that it would speak the language of nature itself. The group of innovators seeking it railed against the tyranny of equal temperament in favor of just intonation, the tyranny of what they felt was a wretchedly narrow European vision of music, the tyranny of the idea of "the scale," and—most of all—the tyranny of the piano. Their hopelessly utopian project, of course, was doomed, as is usually the case with such liberal-minded globalist endeavors. It is a measure of their failure to bind all into one, perhaps, that I have been induced to describe an

Figure 6.4. Prototype twenty-four-key Wheatstone concertina
for "experiment on the formation of the musical scale."
(By kind permission of Neil Wayne and The Concertina Museum.)

Figure 6.5. Robert Howlett's photograph, "Men at Work Beside the Launching Chains of the 'Great Eastern,'" Isle of Dogs, London, November 18, 1857. (The Metropolitan Museum of Art, Gilman Collection, Gift of The Howard Gilman Foundation, 2005.)

anomalous Javanese *metallophone*, strange talking machines, myriad forms of the humble concertina, and neo-Renaissance enharmonic keyboards now gathering dust in underfloor vaults in the museums of London.

THE CIRCULATION OF OBJECTS/
THE CIRCULATION OF KNOWLEDGE

My story begins on the periphery, in the East Indian Archipelago, where, on March 26, 1816, the *Ganges* sailed from the port of Batavia, now Jakarta, on the island of Java. The East Indiaman was laden with upwards of thirty tons of natural and cultural goods. Two hundred "immense packages" were the spoils of the five-year occupation of Sir Stamford Raffles and the British East India Company. Raffles had been recalled to London after the official secession of Java and its dependencies back to Dutch control. The renegade colonialist was in ill health following the death of his wife and failure to convince the court of directors to maintain its "Eastern Insular Empire." Relieved of his post as lieutenant governor, Raffles was charged with commercial oppor-

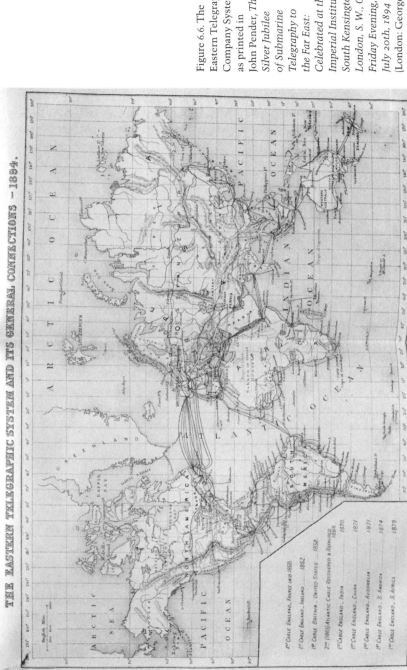

Figure 6.6. The Eastern Telegraph Company System, as printed in John Pender, *The Silver Jubilee of Submarine Telegraphy to the Far East: Celebrated at the Imperial Institute, South Kensington, London, S. W., On Friday Evening, July 20th, 1894* (London: George Tucker, 1894).

tunism, overstepping his authority by annexing Java without official British sanction, and ransacking the royal Javanese court of Yogyakarta. Outlaw or not, historians have noted that Raffles's island sojourn proved the model of later expansionist rule. His administration, Tim Hannigan argues, served as a prototype for high Victorian imperialism. "Knowledge is power," wrote Raffles some four years after his exoneration by the company's court of directors, "and in the intercourse between enlightened and ignorant nations, the former must and will be the rulers."[6]

Raffles certainly stockpiled vast quantities of knowledge in pursuit of free markets and liberal British enlightenment. His getaway secured a vast store of Indonesian antiquities for the East India Company's museum at company headquarters, a gallery free to the London public. The Oriental Repository in Leadenhall Street would later supply the bulk of the impressive South Asia Collections of the Victoria & Albert Museum. In addition, the *Ganges* furnished the source materials for Raffles's lavish two-volume *The History of Java*, published in the year of his landing at Falmouth, and the even more comprehensive three-volume *History of the Indian Archipelago*, penned by British Resident in Yogyakarta, John Crawfurd, who had overseen the seizure of the entirety of the court archives at Raffles's command.[7] The scene onboard the vessel was enclosed again between the pages of the books recording its contents. Javanese space was collapsed, every corner of the island pillaged in order to enrich the vaults of the "Honorable Company." Raffles proved his avarice in February of 1824, when another Indiaman chartered by him, the *Fame*, was destroyed by fire a day's passage from southwest Sumatra, taking down a second and apparently even more comprehensive loot of curiosities: botanical, geological, and zoological.[8] Amongst the *Fame*'s losses, probably, were live specimens for the envisaged gardens of the Zoological Society of London, of which Raffles was founding chairman and president. The *Ganges* made safer passage, landing on July 11, 1816, with John Crawfurd, manuscripts, plants, animal skeletons, Raffles's faithful Malay servant Lewis, weaponry, skins, carvings, geological samples, the naval surgeon and naturalist Dr. Joseph Arnold, specimens of tapir, barking deer, Javanese nobleman and musician Raden Rana Dipura, and nearly ten tons of musical instruments: at least two and perhaps three sets of gamelan.

Once offloaded, this impressive freight spurred the emergence of a host of nineteenth-century comparative disciplines. Raffles's hunger for data, statistics, and curiosities was foundational not only for the pan-European science of comparative zoology but for such liberal humanist endeavors as comparative philology. The future "Father of Singapore" himself was fluent in Malay and a passionate scholar of the religious, institutional, commercial, civic, and literary life of the Indonesian world. Wilhelm von Humboldt admitted his debt to especially Crawfurd in his last and greatest work, the monumental three-volume *Über die Kawi-Sprache* (On the Kavi language), published posthumously in 1836–1839, with an introduction entitled "On Language: The Diversity of Human Language-Structure and Its Influence on the Mental Development of Mankind."[9]

Citing Crawfurd, Humboldt drew together the "fragments of a sacred language now unintelligible to [the Javanese] themselves." Famously, the great Prussian linguist proposed kinships between the sounds of words from a host of dialects in order to establish the truth of what he called the Malayan-Polynesian family of languages. On the basis of sounding signifiers, Humboldt theorized great movements and mental affinities between distant peoples. His theory of cognate languages set the stage for his illustrious student Franz Bopp, who published *Über die Verwandtschaft der malayisch-polynesischen Sprachen mit den indisch-europäischen* (On the kinship of the Malayan-Polynesian language to the Indo-European) in 1841. The spoils of colonial pillage, in other words, equipped a new breed of philologists to study the deep history of phonemes, measure mental difference, map vast population movements, and order the family of nations.

In related ways, the Raffles gamelans furnished raw material for a coterminous comparative mapping of global musical knowledge. These particular instruments became key to the classification of racial difference through the nineteenth century and beyond. It is likely, for example, that William Crotch, sometime professor of music at Oxford University, incorporated his view of Raden Rana Dipura's playing of the *gambang* and the Raffles gamelan that he inspected at the Duke of Somerset's residence into his popular public lectures "On the Music of Ancients and National Music," by which he entertained members of the Royal Institution in May of 1824 and 1829. As quoted by Crawfurd, Crotch theorized a "common enharmonic scale" as the key to "the real native music of Java." In this way, he followed Humboldt's writings on ancient Kawi, where the language's affinities with Sanskrit were distinguished from its Malayan elements. Bizarrely, Crotch drew connections between the Javanese scale and the "scale so many of the Scots and Irish, all the Chinese, and some of the East Indian and North American airs."[10]

Crotch was clearly ignorant of the inauthenticity of the "Indonesian" tunings presented to him. The most notable of Raffles's instruments remain the fabulously oversized zoomorphic gamelan now kept by the British Museum's Department of Ethnology. As Sam Quigley has argued, Raffles probably commissioned these "antiquities" of dragon-and-peacock exotica for visual display in London. A second gamelan—the one inspected by Crotch—now re-

Figure 6.7. The Raffles *gendèr*, now at Claydon House, Middle Claydon,
Buckinghamshire. (By kind permission of Sir Edmund Verney.)

sides in Claydon House in Buckinghamshire, having previously moved from
the Duke of Somerset's on Park Lane to the East India Company's museum,
probably around 1825 (see figure 6.7).[11] Quigley speculates that this gamelan
was also made by special order of Raffles, the components assembled from a
hodgepodge of workshops by artisans from different traditions in and around
the northeastern coastal region of Java. Played today, the Claydon House

set betrays the extent of the concessions made to European diatonic tunings, once again to ease performance and communicability in London. For instance, the strange two-octave *gendèr*, introduced and illustrated by Myles Jackson in his description of Wheatstone earlier in this volume, with its elaborate casework and decorative carvings, boasts a tuning best described, not in terms of *sléndro* norms, but as a European diatonic scale minus the fourth and seventh degrees. The instrument has eleven notes and a strange "butterfly" tuning valve affixed into the two "tonic" resonator tubes, which modifies pitch.[12]

The question of why Quigley should be so obsessed with the exact tone measurement of Indonesian instruments will concern us later, when we explore the long history of measuring Javanese cultural difference. Suffice it to say here that Wheatstone was as practiced at measuring global musical instruments and global scale systems as he was at measuring human skulls, he being an officer of the London Phrenological Association and a lifelong devotee of what (initially at least) was a liberal science. His student Ellis (1885) was a measurer, as were Kunst and Bukofzer (who published measurements in the 1930s and '40s); Jones, Lentz, Hood, and Surjodiningrat (who followed suit in the 1960s); and Rahn (in the late 1970s), to name a few.[13] An illustrious line of comparative and ethnomusicologists measured racial difference in these organological ways, which probably explains at least one reason for the ubiquity of Javanese instruments in music departments throughout the Anglo-American world today. From the beginning, the broken, wide-spaced scales of these objects—particularly as they were played by Raden Rana Dipura—proved a challenge to European norms. (Some sense of Dipura's performances in London can be gleaned from the three rebab or spike fiddle "specimens" reduced to staff notation in Raffles's *History of Java*, and six melodies printed with Crotch's assistance in Crawfurd's *History of the Indian Archipelago*.[14]) Even before they had tackled the problems of notation, the nineteenth-century reformist suspicion was that these Indonesian scalar regimes were just as natural, if not more so, than any European equivalent.

TALKING MACHINES:
DISAGGREGATING SOUND

On the night of Tuesday, March 24, 1835, the same Claydon House *gendèr* reappeared behind the newly erected Corinthian facade of King's College London, having been transported to the Strand from the Oriental Repository. The occasion was the sixth in the "Popular Lectures on Sound" series presented by newly appointed professor of experimental philosophy at the insti-

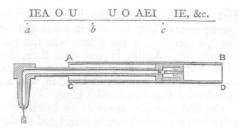

Figure 6.8. Reed pipe, or phthongometer, by Willis, with vowels produced by adjusting the length of the pipe by means of the movable piston. (W. F. Barrett, "Speaking Machines," *Good Words* 19 [1878]: 490. Doe Library, University of California, Berkeley.)

tution, Charles Wheatstone. This was a major King's event, widely advertised in the press probably in order to swell the reputation of the college. King's, after all, would only be granted the right to confer degrees the following year. Before a diverse public, Wheatstone presented for display, besides the Javanese *gendèr*, a metal plate and bow for the demonstration of Chladni figures, diagrams of those same acoustic figures, a Jew's harp, a "Tsing, or Chinese Organ," and a talking machine ("tubes with extensible sides") modeled on Charles Willis's phthongometer (see figure 6.8).[15] The evening's *conversazione* involved breathtaking demonstrations of "communicated vibrations," "simple and multiple resonance," and "the mutual influence of vibrating tongues and columns of air."[16]

Wheatstone had publicized Raffles's *gendèr* before. The instrument builder was a notoriously indistinct orator. His friend Michael Faraday— who had recently described electromagnetic induction—often substituted for him, as on February 15, 1828, when the *gendèr* was heard at the Royal Institution. Whatever the mechanist's verbal insecurities, his experimental apparatuses certainly never lacked for spectacle. Wheatstone's demonstrations made his institution's enthusiasm for popular knowledge plain, as well as its mission for public access, egalitarianism, and the global diffusion of useful science.

This specimen of Asiatic Enlightenment evoked an astonishing sonority, conjuring its "deep rich tone" by the magic of reciprocal resonance. Wheatstone explained its enigma by inserting pasteboards between each of the eleven metal plates (suspended horizontally by two strings) and their corresponding resonating tubes. When the plates were struck in this covered state, the unusually thick bamboo peculiar to Raffles's instrument stopped resonating. The whole fell almost mute. Where once the plates made hardly any sound at all, the "reciprocated" vibrations of each column of air could be moved into and out of the world of the audible merely by opening or closing

the tubes. Wheatstone wrote of these invocations: "I am unaware of any instrument having yet been manufactured in Europe, in which the unisonant
resonances of columns of air have been made available as a means of augmenting the intensity of sounds."[17]

What Wheatstone was trying to achieve with this show of miracles and
wonders was nothing less than to reform his auditor's view of the nature of
sound. Wheatstone's audience was urged to close listening: to a microphonic
level of attention to a vibrational world beyond everyday hearing. (The word
"microphone" is often erroneously said to have been coined by Wheatstone
nearly a decade earlier in relation to a simple nonelectric instrument he devised for "rendering audible the weakest sounds."[18]) His experiment probed
the *gendèr's* mysterious flowerings of sound, its mighty oriental resonators
animated by powerful reverberations at the edge of audibility. Instruments
such as this penetrated the mystical interior. They annihilated existing aural
space by bringing to experience a resonant darkness beyond the superficial
world of immediate perception. And it is significant that Wheatstone felt it
necessary to requisition the riches of Eastern knowledge to make this as yet
unexplored world of sound known.

Wheatstone's demonstrations conjured an awe-inspiring kind of truth.
On the other side of silence was a strange chaos of vibrations, a dense com-
binatorial network of conflicting signals, waves, and overlapping patterns.
Wheatstone urged his auditors to tune into and out of this Asiatic fog. He exhorted them to listen for the roaring silence and to comprehend that a small
piece of that vibrating world might be captured, made audible in a bamboo
pipe, or made visible in the mysterious hieroglyph of a Chladni figure.[19] The
instruments that he selected for display, in other words, attuned his auditors
to the great telephonic cacophony beyond hearing. The sounds of the *gendèr* induced auditors to cultivate a heightened consciousness and a higher
sense of the complex "over there." As Dickson argues in a previous chapter,
they encouraged enraptured listening, an interiorizing style of awe-struck
absorption, of course, with strong religious and educational connotations.
The instruments on show, in this sense, honored the devotional practices of
the improving classes, whether in relation to sermons in church or instrumental music in emerging concert halls.

The claim of the theory of "multiple resonance" was that sounds were
compiled of many coexistent sounds. This was the gist of Wheatstone's finding that the *gendèr's* resonators could be summoned merely by animating a
gnostic world of vibration external to them. Wheatstone's experiments had
the effect of amplifying this sense: that one tone carried with it an array of
properties, intervals and potential alignments, a whole infinity of partials

that might be isolated and brought within earshot only by being reciprocated in some way. Any perception of "simple sound," in this world, was only the mental effect of our hearing a fractured constellation of pitches simultaneously. The earth itself was alive with sound.

DISAGGREGATING VOICE

In his discussion of "the mutual influence of vibrating tongues and columns of air," Wheatstone presented the Jew's harp, Chinese *sheng*, and phthongometer. All three were free reed instruments. That is to say, each was activated by what Wheatstone called "tongue oscillation," in ways (he was quick to point out) still poorly understood in Europe. The sounds of these air-activated metal or thin ivory reeds, in other words, were reciprocated by associated resonating cavities along the lines of the Javanese *gendèr*. The "vibrating tongue," free on one end and fixed on the other, produced sound by what Wheatstone called "periodical intermittences of the current of air." This system closely matched the action of the human voice, at least according to the fashionable French theories of Jean-Baptiste Biot.[20] All three of Wheatstone's final demonstration apparatuses, in other words, were speaking machines from various parts of the world, as the scientist saw it, engineered to study the voweled qualities of the human voice.

Wheatstone's use of that word "tongue" betrayed that the scientist himself was working on his own vowel machine with a flexible resonator or "mouth" made of Indian rubber in 1835, which he would present to the British Association in Dublin on August 14. The word also betrayed Wheatstone's debt to Christian Gottlieb Kratzenstein, who had famously used a "Chinese" *anche libre* as the "voice source" for his famous 1789 vowel synthesizer. (This instrument famously won the competition, organized by Leonard Euler and the Imperial Academy of St Petersburg in 1779, to identify the true nature of the Latin vowels A, E, I, O, and U.) Kratzenstein had theorized, bizarrely, that the epiglottis was the voice's true sound source. Wheatstone quoted Wolfgang von Kempelen, that other great illusionist and inventor of a famous 1791 speaking machine, to emphasize the level of musicianship required to play languages: "In the space of three weeks," says von Kempelen, "anyone may acquire wonderful skilfulness [*sic*] in performing on the speaking machine, especially if he applies himself to the Latin, French, and Italian languages; for the German language is much more difficult."[21] At this stage in his career, Wheatstone felt that the future of telecommunications lay in perfecting such free-reed technologies, and in perfecting an instrument capable of speaking the sounds of every human tongue. If it were conceivable that musical sound might be conducted from London to Aberdeen, Wheatstone reasoned in 1831, one "way of transmitting speech to a distance might be to use a talking machine."[22]

As an aside, it is tempting to link a prevailing fetish for "tongues" in 1830s voice science to widespread obsessions with tongues in elite European vocal practice of this period. Witness Francesco Bennati, house physician at the Théâtre Italien in Paris, who in 1832 credited the sonorous sounds made by Catalani, Lablache and Santini to the size and shape of their tongues. Reed theories for voice were all the rage in 1830s science, supplanting ancient ideas that the vocal organ resembled a wind or string instrument. In 1831, famously, the fashion for "reedy" singing was imported to London by French soprano Henriette Méric-Lalande, whose pulsating "tremolo" was the subject of much debate. Méric-Lalande's debuted in Vincenzo Bellini's *Il pirata* as Imogene, a role written for her, alongside Giovanni Battista Rubini (the original Gualtiero), leading tenor of the 1830s, whose vocal style was equally notorious for its intense and reed-like tremulousness.[23]

The phthongometer on display in Wheatstone's lecture theater was hardly as eloquent. Willis's instrument was a precursor of the better-known electromagnetic vowel synthesizer built by Hermann von Helmholtz, the German who would construct an instrument that, in more than just the physical sense, resembled Raffles's *gendèr*. Helmholtz would famously align electromagnetically vibrated tuning forks tuned to a harmonic series with a line of reciprocating cylindrical resonators, which could be closed or opened (using piano keys) in the style of Wheatstone's Javanese display. For Wheatstone, the lesson of Willis's phthongometer was that pure vowel sounds were the effect of "multiple resonance," a phenomenon only faintly understood before its construction.[24]

In 1830, Willis reported inserting a free reed into a cylinder with a simple slide for shortening or extending the length of the tube. The professor of mechanics at Cambridge University concluded that a whole spectrum of vowels could be reproduced as the slide was drawn in. Wheatstone heard "IEAOU-UOAEI-IEAOU" as he shortened his version of the tube. Having uncovered the secrets of the Eastern *gendèr*, Wheatstone urged that vowels were merely the effect of partials within any compound sound being isolated and amplified. The trick was, not so much to make sounds, as to align them with an already vibrating universe.

A precise knowledge of the true measure of vowels promised to assist philanthropists in reforming orthographical standards in the name of a rationalized international phonetic alphabet. Extolling the value of talking machines for the *Westminster Review* in 1837 (edited by radical politician and

musical instrument maker Thomas Perronet Thompson), Wheatstone lamented "the extreme inadequacy of our written language" in the representation of vowels, and the "want of correspondence between the characters
of our written and the sounds of our spoken language." "We have six characters which are called vowels," he complained, "each of which represents a
variety of sounds quite distinct from each other."[25] The professor's lament
echoed that of his friend John Herschel, celebrated astronomer and scientist
of sound, who himself followed in the early-century footsteps of Constantin-
François de Chassebœuf. Earlier in the decade, Herschel had bemoaned the
contingency of English letters and looked forward to a worldwide "phonetic
alphabet." His proposition was an alphabet amenable to every tongue, where
"every known language might properly be reduced to writing and pronunciation, which would be one of the most valuable acquisitions, not only to philologists, but to mankind, facilitating the intercourse between nations, and
laying the foundation of the first step toward a universal language."[26] The
liberal fantasy, in other words, was a uniform system of phonetic representation for speech sounds, one that would in fact spawn a new discipline and
a new word coined in 1841—"phonetics," a science that moved against Babel.

Herschel's letters of the period show that he was at the center of a small
circle of reformist thinkers, including Wheatstone, who sought the restitu-
tion of a perfect prelapsarian protolanguage. These Christian socialists, missionaries, and Victorian liberals pursued new landscapes of listening with
evangelical zeal. Principal among them was Christian von Bunsen, Prussian
diplomat and intimate of Raffles, who had inspected the same *gendèr* Wheatstone had when he visited Lady Raffles in March 1839. In his diary, Bunsen
described "the former Queen of the East amidst her relics, and surrounded by
the remains of her station," noting that "her set of Japanese [*sic*] instruments
of music—(plates of brass, &c.) have no *quart* or *septima*, but otherwise our
scale."[27] In early 1854, the diplomat and linguist convened a conference at
his London residence titled the "Universal Missionary Alphabet." His guests
included Herschel, Wheatstone, Charles Trevelyan, Richard Owen, Max
Müller, and Henry Venn, together with representatives from the Asiatic and
Ethnological Societies and "all the great Protestant missionary enterprises."
Conspicuously absent from the group was Wheatstone's friend and disciple,
Alexander J. Ellis (of whom more later). Ellis had invented—in competition
with many other global writing systems of this period—an orthography so
comprehensive as to encompass the sounds of every known dialect. He de-
scribed that system in his *Alphabet of Nature, or, Contributions towards a
More Accurate Analysis and Symbolization of Spoken Sounds; with Some
Account of the Principal Phonetic Alphabets Hitherto Proposed* of 1845.[28]

According to the *Times*, Bunsen's opening address stressed the necessity of deploying the phonological discoveries of comparative philology in light of "the great Protestant missionary movement all over the globe." The holy grail was a rationalized phonetic script. (Victory would have to wait until 1888 and the institution of the International Phonetic Alphabet.) As things stood, the vagaries of regional and English orthography were chaotic. A system of natural spelling was thus urgently required. Not only would a scientific measure potentially improve language learning and literacy for lower class and non-native speakers throughout the empire. A fixing of true vowels would also be of untold pedagogical value in comparative contexts, assisting missionaries in translation, and aiding international intercourse and industry. One application that Willis had long since proposed for his phthongometer was to "furnish philologists with the correct measure of difference in natural vowels" and thus to accurately tabulate mental divergence between world languages.[29] Wheatstone viewed his own talking machine differently: less as a precision instrument for comparative research than as a pedagogical tool for the purification of elocution more broadly. His project, simply put, was to obliterate local accent. It was to harness the instrument as a means to "fix and perpetuate the [uniform] pronunciation of different languages."[30]

SINGING IN TONGUES

Wheatstone's talking machine spawned many new generation imperial communication technologies. Joseph Faber contacted Wheatstone in 1839 while building his Euphonia, a speaking automaton capable of multilingual speech and song ("God Save the Queen" for example), which was exhibited at the Egyptian Hall in the mid-1840s.[31] A young Alexander Graham Bell was so mesmerized by the performance of Wheatstone's "Philosophical Instrument" in 1861 that he set out to create his own speaking machine. Bell's experiments, conducted partly in consultation with Alexander J. Ellis, would eventually lead to the 1876 patent of his "musical telegraph," or telephone, which generated its own formal elocutionary standard—its "telephone manner."

A fourth free-reed instrument on the table for Wheatstone's "Popular Lectures on Sound" was the concertina, the name of which first occurred in his 1835 notes. This bellows-driven speaking machine was multireeded only because of the fact that the single reed of Wheatstone's existing machine only "spoke" in dull monotone. That is, this portable "transport instrument"—its free reeds set into a polygonal plate—was intended to speak in many tongues. Evidence suggests, actually, that in the early days, Wheatstone experimented with ways to make his concertina speak in vowels, using resonators, though these efforts apparently failed. An 1844 patent tendered by Wheatstone for concertina improvements described several hybrid forms, including one clearly meant to alleviate the problem that this talking machine lacked facility in vowels. This involved "a means of modifying and ameliorating the tones of freely vibrating tongues or springs by placing resonant tubes over them."[32] A double-reeded concertina built on this model with acoustically linked reed chambers—a speaking instrument—survives in the Wheatstone Collection archived at King's College London.[33]

Wheatstone's concertina, in other words, was originally a species of talking machine. It was a voice in a box. From the beginning, this acoustic demonstration device was hailed for its "talking qualities" even when it briefly escaped the laboratory and entered the midcentury Victorian concert hall. Having heard the Italian virtuoso most responsible for its strange midcentury incarnation as a vehicle of highbrow solo display, English critic Henry Chorley wrote that the concertina possessed "varieties of tone outnumbering those of any wind instrument,—and besides these, a certain talking quality of voice." The concertina made voices that encompassed the expressive capacities of concerted instruments *tout court*. The *Musical World* reviewed the same Annual Morning Concert on June 22, 1854, lauding the performance of Bernhard Molique's G-major concerto for the "sentiment and expression by which [Giulio Regondi] assimilates his instrument to the human voice, and *sings* in a manner to rival the effects of the greatest singers." The reviewer continued: "The cantabile passages remind us, by their breadth of tone and deep feeling, of Rubini."[34] By 1854, Wheatstone & Co. had taken to manufacturing concertinas according to newly emergent concepts of "voice type." As Charles Dickens noted six months previously, one could speak in any vocal register: treble concertina, baritone concertina, tenor concertina, and so on.[35]

The assemblage that we call "the concertina" was slow to gather. Even as the object was commoditized, prototypes were being tampered with and rebuilt into a bewildering array of related entities. The first trial concertinas appeared in many different patents, and various shapes and sizes: octagonal, hexagonal, square, "Wheatstone," "Anglo," "Duette," "Double," "Table-top," "Foot-powered," "with sliding reed," "Clarionet," with twenty-four, thirty-two, thirty-six, thirty-eight, forty-four, and forty-eight keys in a myriad of fingering systems and tunings. The instrument, according to one midcentury commentator, could "be played in any position, standing, sitting, walking, kneeling, or even lying down."[36] It was at once for research, instruction, and entertainment.

Later versions of Wheatstone's multi-"tongued" reed instrument would be advertised as the sound of "British dominions and Colonies." They were taken to the Antarctic by Shackleton, Central Africa by Livingstone, and were instruments of choice for colonial missionaries. Just four years before their brutal murder on Tierra del Fuego, members of the Patagonian or South American Mission Society greeted the Yaghan peoples of the islands on October 21, 1855, with "a few notes on the concertina." In April of 1881, the *Times* reported that William Noble of the Blue Band and Army and Gospel Temperance Society had travelled to South Africa in order to present a "handsome concertina" to the King of the Zulus, Cetshwayo kaMpande, who was being held as a prisoner of war near Cape Town. When the Portuguese explorer Alexandre de Serpa Pinto met the Sova of the Cussivi in what is now southern Angola, he found him "extremely well dressed, wearing, over a sort of uniform, a cloak of white linen, with a large and handsome kerchief round his neck. His head was covered with a cap of red and black list, and in his had he carried a concertina, out of which he wrung the most painful sounds."[37] In colonial peripheries such as southern Africa, the concertina had many names: the squashbox, *izibambo zika Satan* ("Satan's handles"), or, as Zulu migrant workers named it after a cheaper Italian derivative, the *iBastari*. In South Africa at least—trust South Africans—concertinas were misused, fiddled with, indigenized, altered, and mutilated in ways that messed with the utopian fantasies of such globalist liberals as Wheatstone.

ANNIHILATING THE SCALE

Wheatstone's 1835 lecture notes show that, in purest natal form, "the concertina" was originally used for an "experiment on the formation of the musical scale."[38] Actually, it would be fair to say that he designed the instrument in order to attack the whole concept of the scale. The concertina layout, after all, explored synchronic resonance and the rich harmonics of compound tones, its black-and-white ivories presenting a series arrangement at odds with the graduated keyboard of the piano. Its studded fingerboards exploded the usual linear motions of ascents, descents, steps, half-steps, and runs. Instead, Wheatstone's matrix placed the truth of music in a space of simultaneity. The fingerboards furnished a kind of *Tonnetz*, or "tone network." A constellation of tones was arrayed under the hands of any player moved to navigate this force field of intervallic attractions. The concertina, in short, taught its users the serious work of listening "through and beyond" instead of "up and down."

Take the button layout of the fully chromatic "standard" forty-eight-key and ninety-six-tongued midcentury treble concertina (see figure 6.9). This de-

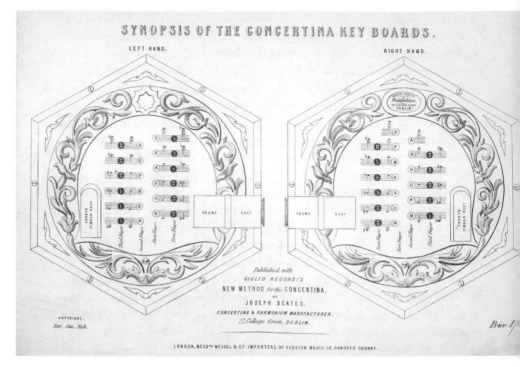

Figure 6.9. Layout on a "standard" forty-eight-button treble Wheatstone, in Regondi,
New Method for the Concertina (Dublin & London: Scates and Wessel, 1857).
(© The *British Library Board*.)

sign placed a symmetrical tonal cosmos of three and a half octaves before the
ear. The pianist's gamut was exploded. The graduated keyboard of the piano
was split in two; the ivories were disaggregated between two fingerboards
separated at either end by cardboard bellows. Wheatstone configured each
button-board as if it were a theoretical diagram of the "universal" principles
of harmony. As Anna Gawboy has expertly observed, the interlaced fifths
and thirds presented an embryonic species of tonal lattice familiar to neo-
Riemannian theorists today.[39] The vertical fifth and horizontal thirds clus-
ters allowed concertists to explore a "table of relations," of the kind later
adopted in midcentury German theoretical texts by Moritz Hauptmann,
Arthur von Oettingen, and Hugo Riemann. The triadic third-fifth scheme
set out nothing less than that curious nineteenth-century invention: "tonal
space." Rapid scale passages, of increasingly questionable "musical" value
anyway, were mind-bogglingly difficult to play, unless of course you were a
"tasteless" Italian virtuoso like Regondi.

Gawboy does not go so far as to suggest that music theorists in the tradition of Hauptmann were closet concertinists. But she does elucidate the common heritage of neo-Riemannian theory and Wheatstone concertinas in the tuning systems of eighteenth-century mathematician Leonhard Euler, who famously published a tuning lattice in order to represent mathematical relations between pure intervals in just intonation. As if in homage to Euler, Wheatstone configured his earliest concertinas around the just-tuned scale of C major. This setup proved perfect for acoustic demonstrations in the major and minor scales of C and E, but less so for such remote keys as B♭ and D major. Wheatstone's concertina, to be clear, furnished seven (rather than five) accidentals to the octave: A♭, E♭, B♭, F♯, C♯, G♯ and D♯.[40] Its two additional buttons allowed for extra fingering choices and enharmonic options at E♭–D♯ and A♭–G♯. Berlioz was mystified when he inspected the lattice for the Great Exhibition in 1851. "The maker of the English concertina," Berlioz wrote, "has introduced enharmonic intervals between A♭ and G♯ and between E♭ and D♯ in the lower three octaves, making A♭ a bit higher than G♯, and E♭ a bit higher than D♯." (According to Ellis, equal temperament only become a "trade usage" in the English piano industry around 1846, and for organs about a decade later.) Berlioz was no fan of "theoreticians" denying the march of equal temperament. He found it "annoying," "absurd" and "barbaric" that flats should be tuned above sharps in the old manner, especially given how progressive musicians now sharpened leading tones in performance. As Wheatstone himself summed up in a presentation to the Royal Society in 1864: "The concertina, invented by Prof. Wheatstone, F.R.S., has fourteen manuals to the octave, which were originally tuned thus, as an extension of Euler's 12-tone scheme."[41]

If Wheatstone blew his own trumpet here, it was only because he was presenting on behalf of Alexander J. Ellis. In fact, the elder scientist was introducing his young colleague as a newly inducted member of the Society. In contradistinction to Berlioz, Ellis was a zealous advocate not only of the phonetic alphabet and just intonation but also of Wheatstone concertinas and John Curwen's Tonic Sol-Fa College, Curwen being the author of that music instructional system eventually rolled out as far afield as the Australian colonies, New Zealand, South Africa, Canada, India, Madagascar, China, Japan, and the South Sea Islands. Furthermore, Ellis was "the true founder of comparative scientific musicology," at least in the mind of celebrated proto-ethnomusicologist Erich von Hornbostel, who declared as much in 1922.[42] The phonetician—another scientist who inspected the Claydon House *gendèr*, achieved his impressive posthumous reputation, apparently, because of the precision instruments that he engineered in order to measure cultural difference. From an early age, Ellis followed Wheatstone in making public exhibitions of a plethora of scalar regimes on an army of variously tuned concertinas. One of his motivations, apparently, was to work against the "unnatural" European drive to equal temperament. Another was to demonstrate the truth, not only of the global diversification of languages, but also of the global diversification of scales. As we have seen, Ellis's search for "the True Scale" would branch into a form of comparative organology. His phalanx of concertinas, in other words, became (for him) a means to capture "the various scales of the nations," and contribute to the measurement of man.

Toward the end of his life, on March 25, 1885, Ellis presented a lecture at the Society of Arts that typified his long-established modus operandi. Ellis's practice, once an indigenous scalar regime had been determined, was to store that scale on one of his myriad concertinas. By the end of his life, Ellis could boast a whole United Nations of concertinas on his shelves, any of which he could open up and compare in public with what he called the "One True or Just Scale of Nature." At this particular lecture, Ellis presented no less than five concertinas, custom-built for him by one of Wheatstone's former business associates. These instruments were tuned to "Meantone" ("the old unequal temperament with extra A flat and D sharp"), equal temperament (which Ellis orientalized by calling the "Meshaqah Arabic scale"!), "Just scale," "Pythagorian," and "Javese," where the white keys played the "Salendro" and the black keys took the "Pelog."[43] In order to capture "Javese" tunings, Ellis relied—predictably—on the same Raffles gamelan that Wheatstone had studied more than half a century earlier. (The aging gamelan had been deposited on loan for that year's display of musical instruments at the International Inventions Exhibition in South Kensington.) The inauthentic "Crotch" tunings (Crotch was a fan of equal temperament) apparently wrecked attempts to analyze the same *gendèr* that had so fascinated Wheatstone more than fifty years earlier. This particular "metal harmonicon," he complained, was "hopelessly out of order, and could not be got to act."[44]

Ellis's strategy was to approach the Indonesian instruments available to him armed with a vast collection of precisely-engineered tuning forks. With these forks, almost a decade earlier, he had devised his celebrated "cents system," a system which would later be foundational to the objective pretensions of early comparative musicology. Ellis's unit of tone measurement—which he unveiled at a meeting of the London Musical Association on November 6, 1876—allowed for unprecedented accuracy in scientific comparisons of "extra-European" scales.[45] In this system, intervals existing between the notes of any fixed-tone instrument found in London's imperial museums or exhibition halls could be calculated down to the hundredths of an equal-tempered semitone. So far as Ellis was concerned, the concertina was perfectly suited to gathering and then performing such pitch collections as the "Javese." As the Cambridge geologist W. Stephen Mitchell noted three years before Ellis's lecture (after studying a troupe of travelling Javanese musicians, whom he heard at the Royal Aquarium recently built to the west of Westminster Abbey): "writers on oriental music have probably dwelt too much on the peculiarity of scales, for in the case of those Javanese instruments, notes in sequence seem never to be used."[46]

At the end of the 1885 lecture, Ellis agreed that the existence of a scaleless Javanese music was proof of the error of Western assumptions. The lesson of Indonesian performance was that scales—all scales—were unnatural. Thus the famous assertion by which Ellis concluded his 1885 lecture: "the Musical Scale is not one, not 'natural,' nor even founded necessarily on the laws of the constitution of musical sound, so beautifully worked out by Helmholtz, but very diverse, very artificial, and very capricious."[47] This statement should not be misinterpreted. Ellis's view was that any collection of tones was necessarily "based on a relation of harmonies, not scales." It was not that the scales of European nations were better than others. Rather, the very concept of the scale was bunk.

No scale was more specious than the European tempered scale. At a March 14, 1877, lecture, Ellis argued this point most forcefully, where (as usual) he played multiple concertinas, this time in "Greek," "Old," "New," and "Just" tuning. This event, presented before the College of Preceptors, was at once a search for the "true basis of music," and an exploration of why scales were so thick with cultural imperfection. "Different nations have chosen very differently," Ellis observed, "and our own present choice seems remarkably strange, and is, at any rate, very modern and extremely artificial."[48] One late-century writer put it well when he discussed the "crystallization of harmonic custom" in "Javanese music-drama," which he found equivalent to the "infinite melos" so heralded by Richard Wagner. "Referable to no rule of our own art," the enthusiast wrote of the gamelan performing twice daily in the summer of 1882 at the aquarium, "these [harmonic laws], by their systematic recurrence point clearly to a rule somewhere, and, though strange, convey an idea that Western music does not embrace every possible good."[49]

The European error, in other words, was "scale thinking." And the injunction was to resist this tyranny with all energy. When the *Musical Times* came to review the second English edition of Helmholtz's *On the Sensations of Tone, as a Physiological Basis for the Theory of Music*, it gasped at the "vanity" of the copious appendices inserted by Ellis, who was also the book's translator. In an extended commentary on "nonharmonic scales," the writer noted, Ellis "asserts that there can be no such thing as a 'natural scale,' but he rather leads his readers to the conclusion that there is no such thing as a scale at all."[50] Writing a year after "On the Musical Scale of Various Nations," the reviewer mocked the "sad insanity" and "intellectual delusion" of Ellis's ten-year-old claim that Helmholtz had "sounded the knell of equal temperament." Despite the lobby for the abolition of temperament, the *Musical Times* observed, musicians were just as "keyboardish" as ever, scientific protests against "German 'pianism' and keyboard theories" notwithstanding. The reviewer found it hard to believe that the greatest musicians of modern times—Bach, Handel, Mendelssohn—were ignorant of "real music." "Is music identical with the piano?" Ellis had insisted in 1877, "Surely there was music in the world before it dreamed of pianos." He continued: "I am afraid we are too apt to identify music with the sound elicited from the piano."[51]

The *Musical Times* was more forceful still in arguing that these "scientific men" had hardly freed themselves from instrumental contingency. If the monopoly of the piano had spawned the vice of "scale thinking," the writer scoffed, then these purported idealists had formed their partial view of music on the basis of the most "faddish" instrument of them all: the enharmonic keyboard. The *Musical Times* had a point. If there was an instrumental template for both the *Tonnetz* of late-century German music theory and the button-board layout of the Wheatstone concertina, it was the new breed of nineteenth-century Ur-Instrument (figure 6.10). The search for pure resonance or the "worship of sensation," as the *Musical Times* put it, had long been conducted in conjunction with the invention of evermore complicated neo-Renaissance keyboards. As the builders of fabulous organs and free-reed harmoniums sought to provide for true intonation in every key, so the standard layout of the piano was dissembled. Keys were split and manuals proliferated, as spaces of harmonic purity were extended before the player. The scale itself came under attack.

Ellis himself ordered the construction of a free-reed "harmonical" that did violence to ordinary sequence. This instrument, on the model of Wheatstone & Co.'s popular free-reed harmoniums, obliterated the scale by inserting red and yellow keys of various shapes and sizes into the octave. More obviously derived from Euler's tuning matrix (and anticipating the concertina's lattice) was an instrument not shown in figure 6.10: the pioneering "euharmonic organ" built by the Reverend Robert Liston, promulgator of a near-just tuning system he called the "Victoria Scale." An early prototype of the euharmonic organ was installed in St. Andrew's Scottish Presbyterian Church in Calcutta from around 1820 to 1857. Another much-discussed instrument was the multicolored enharmonic organ invented by Thomas Perronet Thompson, former governor of Sierra Leone and radical parliamen-

Fig. 6.10. Robert Bosanquet's generalized fingerboard.
(Science Museum, London/Science & Society Picture Library.)

tary reformer who worked as tirelessly for the abolition of temperament as he did for the abolition of slavery. Thompson's broken-scaled organ, with its "quarrils," "flutals," and twenty-nine notes to the octave, was gifted to John Curwen and stood for the latter half of the nineteenth century at Tonic Sol-fa Central in Jewin Street Independent Chapel in London.[52] Under the Reverend Curwen, this near-just instrument trained generations of both metropolitan and colonial singers to maximize natural resonance, and to sing in accordance with the impossible dream of pure harmonic tunings. In South Africa, where Curwen reigned supreme, it is well known that pianists sound bad when they accompany *amakwaya* tonic solfa-ists; and the reason has to do with those tempered pianos of course, which are so miserably out of tune.

When Ellis did recommend "playing" pianos, it was only to turn the instrument into a demonstration device on the model of Wheatstone's experiments with the Javanese *gendèr*. In his 1871 address to the Tonic Sol-Fa College titled "Pronunciation in Singing," Ellis urged his auditors to explore a strange kind of extended keyboard technique. Vocalists, he announced, would do well to raise the dampers of their pianos and sing any vowel loudly and suddenly into the strings. The instrument, it would then be found, echoed back this sound perfectly, proving the basis of vowels in compound harmonic resonance.[53]

In the end, Ellis's ultimate instrument was a "microscopic pianoforte," an impossible apparatus as perfect as the human ear itself. Also at the Musical Association meeting, Ellis exhorted his interlocutors to imagine a "monster pianoforte" with "a keyboard 480 feet long, containing 14,401 strings."[54] Ten years later, at the Society of the Arts, he would dream of "a piano made of such a gigantic size that we could interpose 99 smaller finer keys, between any two at present existing, and that we could tune these at exactly equal intervals, called *cents*, so that 100 cents would form an equal semitone."[55] The strings of this piano, he explained, would act exactly on the model of the "16,400 capillary nerve fibres" of the cochlea in the internal ear. The inner ear, in the words of his address to the newly-founded Musical Association five years later, "may be compared to a microscopic pianoforte with about 16,400 strings tuned to different pitches."

This turning inward, to the realms of a gigantic piano within, brings us to a perhaps inevitable conclusion. It is probably right that the search for the most politically efficacious instrument of them all should fold in on itself in this way—a great internalization at the terminus of a hopeless imperial adventure. This monstrous inner ear reminds of yet another of those compressed spaces so beloved of Victorians, spaces such as the Oriental Repository of the British East India Company, Wheatstone's ecstatic world of telephonic vibration, the "tonal spaces" of concertinas, scaleless enharmonic harmoniums and organs, and—in the end—the generalized space of empire itself. I will conclude by observing that the scientific project to obliterate difference, dialect, scale, and accent can never be violent enough, no matter the appeal to abstraction. The competition over land, for mobility, over the means to bind place to place, is often also a competition over instruments. Those who have the implements and the means to control instrumental standards are also often those who have the power to connect, to separate, and to move the world in ways that work for them.

NOTES

1. For a classic communication studies approach, which I am reversing by treating technology as an organologist would musical instruments, see Jonathan Sterne, *The Audible Past: Cultural Origins of Sound Reproduction* (Durham, NC: Duke University Press, 2003).

2. "The Gatherer," *The Mirror of Literature, Amusement and Instruction* 28 (December 17, 1836): 416.

3. A letter from Charles Lyell to John Herschel dated May 26, 1837, reported on the hype around "Wheatstone's new plan for telegraphing information" in the company of Charles Babbage the previous evening. Wheatstone's first telegraph apparently used five wires, as in the musical stave, each line "representing a letter of the alphabet or several letters, [indicated] I believe by different intensities of the charge." On June 21, Lyell wrote again on the possibility of a "rope in the sea." Katherine Murray (Horner) Lyell, *Life, Letters and Journals of Sir Charles Lyell*, 2 vols. (London: J. Murray, 1881), 2, 13.

4. See Iwan Rhys Morus "To Annihilate Time and Space: The Invention of the Telegraph," ch. 7 in *Frankenstein's Children: Electricity, Exhibition, and Experiment in Early Nineteenth-Century London* (Princeton, NJ: Princeton University Press, 1998), 194–230.

5. Isabelle Stengers, *Cosmopolitics*, trans. Robert Bononno, 2 vols. (Minneapolis: University of Minnesota Press, 2010).

6. In a letter of October 9, 1820; Lady Sophia Raffles, *Memoir of the Life and Public Services of Sir Thomas Stamford Raffles* (London: James Duncan, 1830), 478; Tim Hannigan, "When Raffles Ran Java," *History Today* 61, no. 9 (September 2011): 10.

7. Thomas Stamford Raffles, *The History of Java*, 2 vols. (London: Black, Parbury & Allen, and John Murray, 1817); John Crawford, *History of the Indian Archipelago. Containing an Account of the Manners, Arts, Languages, Religions, Institutions, and Commerce of its Inhabitants*, 3 vols. (Edinburgh: A. Constable and Co., 1820).

8. "Loss of the Ship Fame," *Singapore Chronicle*, April 29, 1824.

9. Wilhelm von Humboldt, *Über die Kawi-sprache auf der Insel Java, nebst einer Einleitung über die Verschiedenheit des menschlichen Sprachbaues und ihren Einfluss auf die geistige Entwickelung des Menschengeschlechts*, 3 vols. (Berlin: Dümmler, 1836–1839).

10. Crawfurd, *History of the Indian Archipelago*, 1:339.

11. The zoomorphic set was kept at Raffles's country manor at High Wood, Hendon, Middlesex from mid-1825. Sam Quigley, "The Raffles Gamelan at Claydon House," *Journal of the American Musical Instrument Society* 22 (1996): 5–41.

12. Ibid.

13. A historical genealogy of tone measurers in the tradition of Ellis is presented in Roger Vetter's "A Retrospect on a Century of Gamelan Tone Measurements," *Ethnomusicology* 33, no. 2 (1989): 217–27.

14. This notation is expertly analyzed in Benjamin Brinner, "A Musical Time Capsule from Java," *Journal of the American Musicological* Society 46, no. 2 (1993): 221–60.

15. William Whewell used the word "phthongometer" in his description of the instrument built by Robert Willis, fellow of Gonville & Caius College Cambridge; Whewell, *History of the Inductive Sciences*, 3 vols. (London: J. W. Parker, 1837), 2, 336. Willis described his instrument in "On the Vowel Sounds, and on Reed Organ-Pipes," *Transactions of the Cambridge Philosophical Society* 3 (1830): 231–68.

16. "Lectures on Sound," *Supplement to the Musical Library* 19 (October 1835): 101.

17. Wheatstone explained that the resonators acted "to augment, I may say to render audible, the sounds of the vibrating metallic plates"; "On the Resonances, or Reciprocated Vibrations of Columns of Air," *Quarterly Journal of Science, Literature, and Art* 3 (March 1828): 175–83, 179.

18. Charles Wheatstone, "Experiments on Audition," *The Quarterly Journal of Science, Literature and Art* 24 (July–December 1827): 67–72, 70.

19. For more on Wheatstone's conception of the world of sound, see the introduction to John Picker's *Victorian Soundscapes* (Oxford: Oxford University Press, 2003).

20. Jean-Baptiste Biot, *Précis élémentaire de physique experimentale*, 2 vols. (Paris: Déterville, 1817), 1:399.

21. Charles Wheatstone, "Review Article," *London and Westminster Review*, October 1837, 21.

22. Brian Bowers, *Sir Charles Wheatstone FRS: 1802–1875* (London: H.M.S.O., 2001), 34.

23. "Madame Lalande possesses a voice of sufficient but not remarkable compass, the quality of which is rather full though reedy in the middle part of it, but wiry and harsh above. It is tremulous to a distressing degree, exciting the idea of age or infirmity, and delivered in the manner of the old French school, which always inculcated the necessity of displaying the powers of the chest to their fullest extent" ("The Drama," *The Harmonicon* 8 [1830]: 221).

24. For a superb study of Helmholtz, the history of vowels, and his English reception, see Benjamin Steege, *Helmholtz and the Modern Listener* (Cambridge: Cambridge University Press, 2012).

25. Wheatstone, "Review Article," 17 (original emphasis).

26. John Herschel, "Sound," *Encyclopædia Metropolitana, or, Universal Dictionary of Knowledge*, ed. Edward Smedley (London: Printed for Rest Fenner, 1830), 2:818.

27. Frances Bunsen, *A Memoir of Baron Bunsen, Late Minister Plenipotentiary and Envoy Extraordinary of His Majesty*, 2 vols. (London: Longmans, Green & Co., 1868), 1:512.

28. Proposals for a universal phonetic spelling occur in Alexander J. Ellis, *The Alphabet of Nature* (London: S. Bagster; Bath, I. Pitman, 1845).

29. Willis, "On the Vowel Sounds," 243.

30. Wheatstone, "Review Article," 22.

31. "The Euphonia, or Speaking Automaton," *Illustrated London News*, July 25, 1846, 59.

32. Charles Wheatstone, "Concertinas and other Musical Instruments." Patent no. 10,041, Great Britain, 1844.

33. A concertina of this type survives in the Wheatstone Collection, King's College London (catalogue number: C1272). This prototype features several sympathetic resonating chambers that amplify the vibrations of associated reeds. It incorporates reed beds or reed plates with two "tongues" each, probably an early attempt to enhance the volume of the instrument.

34. "Concerts of the Week," *Athenaeum* 1391 (June 24, 1854): 789; "Signor Giulio Regondi," *The Musical World* 25, no. 33 (June 24, 1854): 423.

35. Charles Dickens, "The Harmonious Blacksmith," *Household Words*, December 24, 1853, 402.

36. W. Cawdell, *A Short Account of the English Concertina, its Uses and Capabilities, Facility of Acquirement, and Other Advantages* (London, W. Cawdell, 1865), 13.

37. *Fingering System of the "Wheatstone" Concertina* (London, 1959); "Cetywayo and Temperance," *Times* (London), October 17, 1881; G.W. Phillips, *The Missionary Martyr of Tierra del Fuego, Being the Memoir of J.G. Phillips* (London: Wertheim, Macintosh, and Hunt, 1861), 55; Alexandre de Serpa Pinto, *How I Crossed Africa*, trans. Alfred Elwes, 2 vols. (London: Sampson Low, Marsten, Searle, & Rivington, 1881), 1:319.

38. Neil Wayne, "The Invention and Evolution of the Wheatstone Concertina," *Galpin Society Journal* 62 (2009): 244.

39. Anna Gawboy, "The Wheatstone Concertina and Symmetrical Arrangements of Tonal Space," *Journal of Music Theory* 53, no. 2 (2009): 163–90.

40. Hector Berlioz, *Berlioz's Orchestration Treatise: A Translation and Commentary*, ed. Hugh MacDonald (Cambridge: Cambridge University Press, 2002), 305–11.

41. Alexander J. Ellis, "On the Conditions, Extent, and Realization of a Perfect Musical Scale on Instruments with Fixed Tones," communicated by C. Wheatstone. *Proceedings of the Royal Society of London* 13 (January 7, 1864): 103.

42. For another panegyric in the tradition of Hornbostel, see Jonathan P. J. Stock, "Alexander J. Ellis and His Place in the History of Ethnomusicology," *Ethnomusicology* 51, no. 2 (2007): 306–25.

43. Ellis's "Just" concertina survives as instrument number M9a–1996 in the Horniman Museum, whose pencil marks on this instrument suggest that he was experimenting with different temperaments.

44. Alexander J. Ellis, "Appendix to Mr. Alexander J. Ellis's Paper on 'The Musical Scales of Various Nations,'" *Journal of the Society of Arts* 33 (1885): 1107.

45. For an effective critique of the purported "objectivity" of Ellis's cents system in view of variable gamelan tunings on the ground, see Vetter, "A Retrospect on a Century of Gamelan Tone Measurements," 217–27. Vetter notes the strange enormity of the craze for Javanese tone measurement in the wake of Ellis.

46. W. Stephen Mitchell, "Musical Instruments of the Javanese," *Journal of the Society of Arts* 30 (September 29, 1882): 1019–21.

47. Alexander J. Ellis, "On the Musical Scale of Various Nations," *Journal of the Society of Arts* 33 (1885): 485–527.

48. Alexander J. Ellis, *On the Basis of Music* (London: C.F. Hodgson & Son, 1877), 17.

49. D. T., "Music from the East," *Musical Times*, September 30, 1882, 606–7.

50. "Reviews," *Musical Times*, August 1, 1886, 481–84.

51. Ellis, *On the Basis of Music*, 3.

52. Patrizio Barbieri, *Enharmonic Instruments and Music, 1740–1900* (Latina: Il Levante Libreria Editrice, 2008).

53. Alexander J. Ellis, *Pronunciation for Singers* (London: Curwen, 1888), 11.

54. Alexander J. Ellis, "On the Sensitiveness of the Ear to Pitch and Change of Pitch in Music," *Proceedings of the Musical Association* (November 6, 1876): 1–32, 7.

55. Ellis, "On the Musical Scale of Various Nations," 487.

CHAPTER SEVEN

Good Vibrations: *Frankenstein* on the London Stage

SARAH HIBBERD

Sudden combustion heard, and smoke issues, the door of the laboratory breaks to pieces with a loud crash—red fire within. Music. The Demon discovered at door entrance in smoke, which evaporates—the red flame continues visible. The Demon advances forward, breaks through the balustrade or railing of gallery immediately facing the door of laboratory, jumps on the table beneath, and from thence leaps on the stage, stands in attitude before Frankenstein, who had started up in terror; they gaze for a moment at each other. "The demon corpse to which I have given life!" Music.—The Demon looks at Frankenstein most intently, approaches him with gestures of conciliation. Frankenstein retreats, the Demon pursuing him.[1]

These words appear in the printed text of Richard Brinsley Peake's 1823 stage adaptation of Mary Shelley's *Frankenstein*, which debuted at the English Opera House just off the Strand in July 1823. This "romance" (as the playbills called it)—featuring a monster who communicated through mime rather than speech—stimulated excitement and controversy. "The audience crowd to it, hiss it, hail it, shudder at it, loath it, dream of it, and come again to it," wrote the critic of the *London Magazine*: "The piece has been damned by full houses night after night, but the moment it is withdrawn, the public call it up again—and yearn to tremble before it."[2] The play was picketed, leaflets were exchanged, and reviewers argued about its morality. On one side were those who simply objected to the subject matter: the presumption of man playing God.[3] On another were those who approved of the play as a cautionary tale: "Man cannot pursue objects beyond his obviously prescribed powers, without incurring the penalty of shame and regret at his audacious folly."[4] More intriguingly, several reviewers noted that while Shelley's novel had offended some with its quasiscientific reasoning, Peake's play was able to delight audiences without infringing on good taste: "The modes of reasoning, principles of action, &c. . . . [are] all carefully kept in the background. Nothing but what can please, astonish, and delight, is there suffered to appear."[5] Indeed, for the critic of the *Morning Chronicle*, "Melo-dramatic action [conveyed] what it would have been extremely hazardous to attempt to express by words."[6] In short, Peake's drama—one of at least eight plays on the *Frankenstein* story staged in London's theaters in the 1820s—offered a compelling dramatization of a controversial topic.[7] The widespread appeal of this "melo-dramatic opera" was heightened by the suggestive mix of music and mime,[8] which by then were well established tools of the illegitimate theater.[9]

Shelley's novel, published in 1818, has been viewed as the most important cultural response to a scientific controversy that erupted in 1814 over whether the body is the source, or merely a conductor, of the life-giving force.[10] In his introductory lecture to the Royal College of Surgeons, professor of anatomy John Abernethy explained that electricity (or a similar power) was a "superadded" force that initiated life in an inanimate body.[11] Buttressing his claims with reference to Humphry Davy's work on electricity, he suggested that vitality was a "subtile [*sic*], mobile, invisible substance,"[12] which occupied an intermediary position between the material body and the immaterial mind and emotions. The counterattack came from Abernethy's erstwhile protégé, William Lawrence, who believed firmly that "organisation is the instrument."[13] He openly ridiculed the idea that electricity could do duty for the soul, as "subtle matter is still matter."[14] The property we call life, he argued, was an emergent effect of more or less complex organizations of matter; there was no need to appeal to some extrinsic vital force. He was influenced by the French school of anatomy, especially Étienne Geoffroy Saint-Hilaire, with whom he shared a fascination with the monstrous as "a demonstration of nature's unfathomable and always surprising possibilities of self-transformation, metamorphosis and transmutation."[15] Between 1817 and 1819 Abernethy and Lawrence expressed their views in lectures, encouraged by their student supporters. The debate was personal and vitriolic, stoking up enormous public interest amid wider concerns about the social and political role of public science.[16] Lawrence was vilified as a dangerous and blasphemous radical, a "materialist" whose approach denied the creative role of God; Abernethy evoked British spirit in his defense against (French) philosophical and political radicalism. To believe in a superadded principle of life was to believe in law and order and moral virtue (though Lawrence, of course, thought Abernethy "immoral" and "materialist"); Lawrence's theory was interpreted as a republican one, leveling all matter and denying that a "monarch," or controlling force, was needed.[17] In 1819, Lawrence had to defend himself against the implication that he had "perverted [his] honourable office" as a professor at the Royal College of Surgeons by "propagating opinions detrimental to society."[18] The debate continued, and by 1822 the vitalists seemed to have won the battle—though of course, in the long nineteenth century, Lawrence's increasingly "bourgeois" conception of organic life prevailed, and the vitalists were themselves vilified as "French," "materialist" and "republican."[19]

Shelley's novel has been understood widely as a skeptical (Lawrencian) response to the "superadded" principle of life: Dr. Frankenstein is the "blundering experimenter" employing Abernethy's old-fashioned ideas, and the seriocomic tone of the novel seems to echo Lawrence's mocking manner in his lectures.[20] However, it can be read in other ways. Vitalism offered Shelley a suggestive language for exploring the complex senses in which bodies are animated, fluid and in flux, rather than governed by the supervening control of a rational mind. Theater—and musical theater in particular—provided a productive space in which to bring such ideas to life. I suggest here that Peake's *Frankenstein* melodrama deploys music first as the animating impulse, and then as a "subtile [*sic*], mobile, invisible substance" itself, softening the monster as he discovers his faculties—as he moves from lifelessness to expressive pantomime to the edge of verbal eloquence. His "soul" begins to take shape through a sympathetic exchange of emotions with those around him (including the audience). But when the monster's gestures toward others are rejected, the fragile state of sympathy crumbles, and he lashes out. A cautionary tale, perhaps, but in its stage adaptation *Frankenstein* offered a compelling portrayal of vitalism's potential to rescue the soul from the threat of "materialist" science.

This chapter will consider ways in which materialist and vitalist theories from the late eighteenth century infiltrated popular as well as scientific culture during the subsequent decades. In this context, electricity and music often appeared as complementary or interchangeable forces; they shared an ability to both calm and stimulate the nerves, and they both inhabited the ambiguous space between the material and the intangible. What is more, Peake's *Frankenstein* play, along with another "peculiar romantic, melo-dramatic pantomimic spectacle" on the same subject by Henry Milner, can demonstrate how the monster of Shelley's novel was transformed into a more vital being on the stage: a being whose experience of music intersected with contemporary views about its physiological, associative, and aesthetic effects. Music thus became a tangible, "sensible" expression of the vital life force on the one hand and a vehicle for the monster's emerging sensibility on the other. Far from aligning with the skepticism of Lawrence's "materialist" beliefs, then, the plays thus dramatize vitalism's progressive potential as a dynamic and creative force.

LIFE SCIENCE

Alan Richardson has described a complex web of scientific theories that informed the evolution from a mechanistic or "corporeal" to a biological or "embodied" notion of mind beginning in eighteenth-century Britain. This uneven process saw the emergence of a cautious fascination with the idea of electricity in neural transmission.[21] In the middle of the century, David Hartley, building on the work of Thomas Hobbes and John Locke, had reduced all mental functioning to association, proposing a process of "vibrations" in the brain and nerves as a material explanation for psychological phenomena.[22] For instance, for Hartley, the enjoyment of music was derived partly from the corporeal pleasures of beautiful sound, and partly from associated delights: ideas stimulated by those sounds without the intervention of the will, which affect the emotions of the hearer.[23] By the end of the century, however, scientists began to recognize the mind as not simply a passive receptor of vibrations, but an active participant in human experience; this new understanding was based on a biological conception of physiological and mental functioning.[24] If musical strings had long been a common metaphor for the nerves, now the aeolian harp emerged as a model for this more integrated understanding of the human mind and body: the harp trembled in response to the wind, but it also transformed the force of the wind into harmonious sounds. Shelley Trower has argued that this model of embodied consciousness might serve as a bridge between "classical" and "modern" accounts of sensitivity: vibration stimulates mind and body (both conceived as matter) into life and sentience.[25] Erasmus Darwin employed a series of visual allusions to demonstrate how much work the brain must do to produce the images we see.[26] Crucially for Darwin, the embodied mind was a sensorium consisting of both the nervous system and the spirit of animation—the latter residing throughout the body, recognizable only by its effects, and consisting of "matter of a finer kind" analogous to electricity.[27] Neurological discoveries in the first decades of the nineteenth century identified a basic distinction between sensory and motor nerves, first described by the Scottish surgeon Charles Bell in 1811.[28] For Bell, the mind developed holistically, in and through embodied experience, realizing the "spirit" through a material body. The divide between the materialist theory of vibration first articulated by Hartley and the idea of an animating power that was to underpin the 1814 debate was not as clear-cut as is often suggested. Darwin, for instance, was attacked as a materialist in the 1790s and as a vitalist in the 1810s.[29]

The question of the source of the body's "electricity" was also at the heart of another famously lively dispute at the end of the eighteenth century. In 1791 the Italian physician Luigi Galvani announced that he could produce electricity from animal tissue.[30] He had noticed that the frog legs his wife was preparing for dinner had twitched when lightning struck during a thunderstorm, and subsequently he discovered that the nerve and muscle twitched when connected through a metallic circuit. He argued that the brain was the source of this "animal electricity," which was then conducted through the nerves to the rest of the body and stored in the muscles—although atmospheric disturbance could override the kinetic agency of living matter. His claims were challenged by Alessandro Volta, who declared that the frog legs simply acted as conductors for the electricity produced by the metal strips that completed the circuit. The dispute continued long after Galvani's death in 1798, in part because the topic leant itself to public demonstration: animal electricity could be investigated "wherever frogs were to be found."[31] Galvani's notion of electricity as the means by which nerve cells passed signals to the muscles (bioelectricity) and Volta's establishing of a more reliable source (the battery, or Voltaic pile) together helped speed rapid change in electrical science during the first decades of the nineteenth century.

Galvani's nephew Giovanni Aldini came to London in 1802, where he performed experiments at the Great Windmill Street Anatomical Theatre and in front of the medical students of Guy's and St. Thomas's hospitals: he typically used frogs, decapitated dogs and rabbits in order to persuade his spectators that "the direct production of the galvanic fluid, or electricity, by the direct or independent energy of life in animals, can no longer be doubted."[32] In 1803, he had the opportunity to experiment on a human body at the Royal College of Surgeons, attempting to revive the corpse of the murderer George Forster six hours after he had been hanged at Newgate; he connected "a pile of 120 plates of zinc and copper" to various parts of Forster's anatomy.[33] There were, inevitably, graphic reports made in the press—ones that anticipated, and perhaps informed, Mary Shelley's fictional description:

> On the first application of the process to the face, the jaw of the deceased criminal began to quiver, the adjoining muscles were horribly contorted, and one eye was actually opened. In the subsequent part of the process, the right hand was raised and clenched, and the legs and thighs were set

in motion. It appeared to the uninformed part of the by-standers as if the
wretched man was on the eve of being restored to life.[34]

Aldini was forced to leave England in 1805 following the public clamor
his demonstrations prompted, but others took up where he left off. In 1818
the Glaswegian physician and chemist Andrew Ure reported on a series of
electrical experiments that he had conducted on another executed murderer, Matthew Clydesdale. When electrical charges were passed between
the supraorbital nerve on Clydesdale's forehead and heel, Ure explained:

> Every muscle in his countenance was simultaneously thrown into fearful
> action; rage, horror, despair, anguish, and ghastly smiles, united in their
> hideous expression in the murderer's face, surpassing by far the wildest
> representations of a Fuseli or a Kean. At this period several of the specta-
> tors were forced to leave the apartment from terror or sickness, and one
> gentleman fainted.[35]

As this reference to star actors implied, for many Londoners (and Glas-
wegians) these events were simply another form of public entertainment.
Indeed, Iwan Rhys Morus has characterized such attempted resurrections
as the denouement of performances that began with criminal conviction,
proceeded with brutal public execution and concluded with the dismem-
berment of the corpse.[36] But Aldini and others sought to justify such work
morally by suggesting that electricity had a valuable medical application as
a means of saving the victims of drowning and suffocation or as a cure for
the insane. Less dramatic employment of electricity's restorative powers
had already been recognized by Charles Kite, who in 1785 won the silver
medal of the Humane Society with an essay on the use of electricity in the
diagnosis and resuscitation of persons apparently dead: "Electricity was . . .
applied, and shocks sent through in every possible direction; the muscles
through which the fluid [electricity] passed were thrown into strong con-
tractions."[37] Of course electricity had a tradition of more general therapeutic use, most famously perhaps in Franz Anton Mesmer's demonstrations of
"animal magnetism," which (charges of charlatanism notwithstanding) resonated strongly with the findings of eighteenth-century scientists including
Hartley, Darwin, and Galvani. Mesmer claimed that an electrical energy or
fluid flowed through the nerves, and that sickness—anything from blindness
to general *ennui*—resulted from an obstacle to the flow. Individuals could
restore health and harmony by applying magnets or massaging the body's
"poles."[38]

By the 1820s, reports of electricity in treatments of physical and psychological ailments were ubiquitous in the press. Thus, the *Examiner* recorded that when "as a *dernier resort*" electricity was applied to a child whose body was convulsing after having been thrown down some steps, recovery was complete.[39] A correspondent to the letters page of the *Morning Post* recounted that a young man "who lost his speech suddenly, and continued dumb for eight months" was cured by galvanism, and that the judicious application of electricity led (variously) to a man regaining the use of his legs after prolonged paralysis, a blind man recovering his sight, and a woman with a cancerous breast being saved from the knife. In a postscript the correspondent added that "a Lady who could only swallow while a particular tune was playing on a violin" (and so was obliged to have music at her meals) was also "cured by electricity."[40]

As this last example suggests, music was widely perceived to have mysterious effects akin to those of electricity. Mesmer built on Isaac Newton's notion that "ether" pervaded the atmosphere and thereby offered a medium for influence across a distance: the planets could influence the body "as a musical instrument furnished with several strings, the exact tone resonates which is in unison with a given tone."[41] He argued that animal magnetism could thus be disseminated and reinforced by music, bringing patients into sympathetic relation with their magnetizer.[42] Although mesmerism did not take hold in Britain until the 1830s, reports of its application in continental Europe appeared regularly in the press.[43] Thus, in 1817, the *Literary Panorama* noted "the salutary action of Animal Magnetism and of Music" promoted by the Bolognese doctor Angelo Colò in his recently published treatise.[44] One of Colò's stories described how the nightly seizures of one Signora Cavazzani were cured by a group of musicians, whose playing circumvented the electric rhythms of the seizures, thereby limiting motion to the limbs: "the music cured it with a flash, an electric shock."[45]

Meanwhile, electrification became a common metaphor for the visceral and pervasive effects of music on concert and theater audiences.[46] Giuditta Pasta's voice was "as fascinating to the listener's ear as it [was] electrifying to his soul"; the effect on the audience more generally was like that of an "electric shock, infinitely more flattering than any mere applause, which ran instantaneously round the auditorium."[47] More extravagant and exaggerated stories were regularly published: Stendhal's claim that a Neapolitan physician had counted "more than forty cases of brain fever, or violent convulsions" among the young female audience at Rossini's *Moïse*, in response to "the superb change of tone" in the third-act prayer, was reprinted in the *Family Oracle of Health* in 1824 and the *Athenaeum* in 1828.[48] The terminology for music's physical and associative effects at this time was drawn largely from the eighteenth-century physician John Brown's theory of nervous excitability. For Thomas Beddoes (who had edited Brown's work in the 1790s), for example, music became the potential cause of neuropathological conditions.[49] Peter Lichtenthal, in his widely read *Der musikalische Arzt* (The musical doctor) of 1807, which otherwise promoted a positive view of music's effect, prescribed "doses" to improve certain conditions, claiming "music must necessarily have damaging consequences when the activity of the heart and blood vessels is increased."[50] As a "vitalist" interest in electricity was growing in the public sphere—its power to animate and shock— so fascination was growing with music's capacity to similarly mediate between the material body and the invisible mind.

FRANKENSTEIN

Claims for the powerful effects of electricity and music on the nation's physical and psychological health were predicated on an idea of the body as a conductor of electrical (or similar) forces that stimulated sympathetic vibration. It was the notion of an extrinsic animating power in such circumstances that William Lawrence was so keen to refute. He has received considerable attention among literary scholars for being the friend and physician of the Shelleys, with *Frankenstein* viewed as a work that picks up where he left off—for one commentator, "promoting a materialist psychology . . . that in the wake of [his] 1819 humiliation could be advocated only in the guise of poetry and fiction."[51]

Percy Shelley first encountered Lawrence in 1812 (through Mary's father, William Godwin), and consulted him about his health in 1815; the two men met several times during the period of the notorious lectures at the Royal College of Surgeons.[52] It is likely that Mary participated in their conversations: the Shelleys' reading notes suggest they were both well versed in the medical and scientific literature of the time.[53] Mary began writing *Frankenstein* in 1816, completing it in 1817 at a time when she was also having medical consultations with Lawrence.[54] Richard Holmes has suggested that rather than taking up Lawrence's views, however, Mary offered a corpse dissection in reverse, taking Aldini's demonstrations to an imaginative extreme.[55] Holmes builds on Marilyn Butler's assertion that the novel seems to draw more on "spectator-orientated demonstration" than on books of science.[56] Indeed, he speculates that Dr. Frankenstein was a composite of a whole generation of scientific men—not only Lawrence and Aldini, but Joseph Priestley, Henry Cavendish, and Humphry Davy, as well as contemporary German scientists such as Johann Wilhelm Ritter.

In spite of the links with Lawrence, then, and *Frankenstein*'s reputation as a "materialist" cautionary tale, it seems that the vitalist potential of the story was more compelling for Mary: the original preface boasted that the novel "affords a point of view to the imagination for the delineating of human passions more comprehensive and commanding than any which the ordinary relations of existing events can yield."[57] In the preface to the 1831 edition she went further, suggesting that "galvanism had given token of such things [i.e., the reanimation of a corpse]: perhaps the component parts of a creature might be manufactured, brought together, and endued with a vital warmth."[58] By moving away from a narrow idea of Lawrence's influence on the Shelleys, then, and acknowledging the complex fascination with electricity that prevailed in scientific and popular culture of the period, we can understand the novel as participating more creatively in the scientific debate.

Its theatrical adaptations complicate this picture even further. As we shall see, the musical dramatizations of *Frankenstein* shift the focus away from the moment of animation to the development of the "soul"—the relationship between mind and matter—in a manner rooted not only in scientific writing and practice, but also in the literary and musical sensibilities of the turn of the century. Peake's and Milner's dramas each offered a more arresting representation of the monster's emerging sensibility than did the novel. They dispensed with Shelley's three interlocking narratives—the framing device provided by the polar explorer Walton and the reports of Frankenstein and the monster—in favor of a direct dramatization of the monster's life.[59] In both works music becomes a (palpable) vitalist force, animating the monster, stimulating his sensory awakening, and conveying his emergent sensibility.

These stage *Frankenstein*s derived much of their effectiveness from the fact that the monster was a pantomime role, his movements therefore accompanied by music. At one level, mute action was a standard feature of the minor playhouses, which had traditionally offered a repertory of burlettas, melodramas, and extravaganzas (featuring pantomime, dance, music, and acrobatics) and had been forbidden from staging spoken tragedy and comedy, which was the province of the patent theaters. But recent years had seen the gap close, as the patent theaters were increasingly drawn to the popular appeal of spectacle, and the minor theaters exploited loopholes that enabled them to stage quasilegitimate drama in ingenious combinations of music, dialogue, and gesture.[60] While theater regulations did not prevent the mon-

ster from having a voice, then, the generic heritage, popular appeal, and creative potential offered by musically accompanied mime made mute monsters a popular choice. Mary Shelley attended a performance of Peake's play,
and noted approvingly, "I was much amused, & it appeared to excite a breath-
less eagerness in the audience."[61]

At another level, as Elaine Hadley has demonstrated, by the first decade
of the nineteenth century pantomime had become associated with the lower
ranks of spectators who frequented the minor theaters, and—through the
comic violence of the harlequinade—it had come to represent both the "farce
of state power" and the physical suppression of a voiceless underclass.[62] Con-
versely, during the Old Price riots at Covent Garden in 1809, the readers of the
Riot Act were reduced to gesticulating from the stage in order to be understood through the clamor; this enforced "pantomime" was understood by
commentators to be the result of either mob violence or popular empower-
ment, depending on one's political perspective.[63] The monster's lack of voice
in this charged context is suggestive: Jane Moody has claimed that Milner transformed *Frankenstein* into a "quasi-political drama about rebellion
against an autocratic power,"[64] and other scholars have viewed the monster as
a representative of the newly politicized masses.[65] As we shall see, these mute
monsters on the London stage can be understood as appealing to the sympathies of their audiences in a manner that had strong political connotations.[66]

Richard Brinsley Peake's "romantic drama" (or "melo-dramatic opera")
Presumption; or The Fate of Frankenstein opened at the English Opera House
on July 28, 1823, and enjoyed an initial run of thirty-seven performances.[67]
Music was supplied by one Mr. Watson, "whose labours," according to the
Morning Post, "we often loudly applauded."[68] Alongside a handful of solos,
duets, and choruses, there was music to accompany stage action (principally
that of the monster), and onstage performances based on those mentioned
in Shelley's novel, which were woven into the drama. Scores do not survive,
but musical cues are included in the published text, and the play was reviewed widely in the national press. Henry M. Milner's *Frankenstein; or The
Demon of Switzerland*, opened on August 18, 1823, at the Coburg Theatre,
and although no text seems to have survived, Milner produced another version of the story for the same theater in 1826: *Frankenstein; or The Man and
the Monster!*, which was partly based on a Parisian adaptation of Shelley's
novel from the same year.[69] A less substantial musical score was provided by
Mr. T. Hughes: there are no sung numbers, but some incidental music and
onstage performances are included (and it is likely that there were also extended musical passages to accompany the monster's pantomime scenes). No
scores have survived—we are again reliant on cues in the text—and the play

does not seem to have been reviewed in the national press.[70] The Lyceum had a history of accommodating a variety of popular entertainments before it was rebuilt as the English Opera House in 1816 (employing cutting-edge technology and featuring the "animation" of the dead through phantasmagorias and waxwork shows). By the 1820s the theater had a reputation for making the "popular" respectable for an increasingly bourgeois clientele drawn from across the city.[71] The Coburg, by contrast, was situated on the south bank of the river and dependent on local, working-class audiences; it had a reputation for sensational blood-and-thunder melodramas.[72] The tone of each play was in keeping with the broader repertory of the host theater: there were no onstage murders in Peake's play, which helped to support a relatively sympathetic portrait of the monster; in Milner's drama, by comparison, several killings took place on stage, and the portrait of the monster was more brutal and—in musical terms—less nuanced.

The actors T. P. Cooke and O. Smith played the monsters in Peake's and Milner's plays, respectively. Both men were well known on the London stage for their portrayals of physically powerful but inarticulate outsiders, who were responsive to music—a tradition that also included wild men, clowns, freaks, and sailors.[73] The muteness of the monsters was rooted firmly in this popular tradition, which in turn contributed to their enthusiastic reception. We do not have any contemporary reports of Smith's performance in this role, though the play-text tells us that he was dressed in "close vest and leggings of a very pale yellowish brown, heightened with blue, as if to show the muscles, &c. Greek shirt of very dark brown, broad black leather belt."[74] Reviewers noted Cooke's "bizarre" and "ethereal" appearance: he had long black hair and light blue skin, and wore a cotton tunic and toga that was often discarded to reveal his athletic build (see figure 7.1). His ability "to strike effective poses or attitudes of both body and countenance," and to render a full gamut of emotions through mime alone was praised. It was widely agreed that he was "tremendously appalling."[75]

How did music animate the monster in each of the two plays, and how did this "subtle, mobile, invisible substance" act in cultivating sensory awareness and his capacity to feel emotions? In the early part of the story music offers a "sensible" means by which the monster gains the sympathy of those around him, including the audience, but these delicate relationships quickly collapse as his desire for revenge descends into a destructive rampage. Later, music is implicated in more disturbing ways: the "demon" runs amok in extended tableaux in which his vengeance seems to be not only supported but encouraged by the orchestra—and in Peake's play by the chorus as well. Music becomes a dangerous stimulant, inflaming the monster's pas-

Figure 7.1. Mr. T. P. Cooke in the role of Frankenstein's monster.
Painted by Wageman. Drawn on stone by N. Whittock. (The Carl H. Pforzheimer
Collection of Shelley and His Circle, the New York Public Library,
Astor, Lenox, and Tilden Foundations.)

sions in destructive ways that, as we shall see, were beginning to be explored by physicians and equated with electricity's capacity to shock.

MUSIC AND SYMPATHY

In Shelley's *Frankenstein*, the role that music plays is clearly underpinned by the classical theory of sensation and association derived from Hartley. In volume 2 of the novel, the monster tells Frankenstein his story, describing his first sensory impressions and his observations of (and encounters with) the De Lacey family:

> I was delighted when I first discovered that a pleasant sound, which often saluted my ears, proceeded from the throats of the little winged animals who had often intercepted the light from my eyes. . . . Sometimes I tried to imitate the pleasant songs of the birds but was unable.[76]

Later in the book, he sees its effect on others: "[The old man] played a sweet mournful air which I perceived drew tears from the eyes of his amiable companion," the man's smile of "kindness and affection" in turn produced in the monster "sensations of a peculiar and overpowering nature" Finally, music's effects are felt directly by the monster: "[Safie] played some airs so entrancingly beautiful, that they at once drew tears of sorrow and delight from my eyes."[77] However, we do not witness the monster's reactions: he describes them in a detached manner; the power of his story thus lies predominantly in his eloquent use of language and demonstration of learning, and in his astute reflections on his experiences.

In the stage adaptations, not only are these scenes brought charmingly to life, but music takes on a more pervasive role in the drama, akin to that of the vital force. In Milner's play, for example, the orchestra accompanies the onstage "creation scene" and gives a sense of life flooding into the inert body:

> (*Music*—[Frankenstein] *eagerly lays his hand on the bosom of the figure, as if to discover whether it breathes.*) The breath of life now swells its bosom. (*Music.*) As the cool night breeze plays upon its brow it will awake to sense and motion. (*Music*—*he rolls back the black covering, which discovers a colossal human figure, of a cadaverous livid complexion; it slowly begins to rise, gradually attaining an erect posture, Frankenstein observing with intense anxiety. When it has attained a perpendicular position, and glares its eyes upon him, he starts back with horror.*) . . . Oh, horror! horror! let me fly this dreadful monster of my own creation![78]

In Peake's drama, though this animation scene takes place offstage, music underpins it, representing the storm (perhaps providing the vital spark), and then the stirring of life while Frankenstein and his assistant look on: "*Music expressive of the rising of a storm. . . . Sudden combustion heard, and smoke issues, the door of the laboratory breaks to pieces with a loud crash—red fire within. Music.*"[79] The orchestra accompanies the monster's appearance and his first gestures toward his creator (cited at the head of this chapter), and indeed the monster's every subsequent appearance—however fleeting. Of course one would expect music to accompany scenes of extended action, but here it does more than that: it stands in for the (unseen) moment of animation, thus offering tangible evidence of the vital force.

What is more, both Peake's and Milner's adaptations contain scenes in which the monster himself hears and responds to music: a device of key importance in characterizing the monster (as reviews attested), and determining his relations with others. In both works, as in the novel, the monster is rejected by his creator immediately (and subsequent "gestures of conciliation" offered by the monsters in both plays are rebuffed). James Chandler has explained how Shelley's novel is permeated with eighteenth-century notions of sympathy and sentiment, and views this rejection as key to the chain of events that follows—as does the monster himself.[80] In his *Theory of Moral Sentiments* (1759), Adam Smith argued that our moral judgments are not dependent on an innate moral sense, but are rather routed through our imagined sympathy with other points of view. Bereft of sympathetic exchange with the people around him and isolated from society, the monster's emergent moral deformity is inevitable. A key premise in the theory of sentiments is that a case of misfortune calls forth a response from the sufferer, which in turn will affect the response of witness to the entire spectacle. Smith explains that restraint in response to unlucky chance will increase the likelihood of a spectator's active sympathy with the victim.[81]

In the plays, the monster's behavior helps to earn the audience's sympathy—a process aided at every step by onstage music. At first appearance, Peake's monster is impressively unformed and unappealing: "a mass of moving matter, without stimulus or intellect—he seems to have eyesight without vision—he moves as if unconscious that he is moving. . . . What can be more harassing than the respiration which supplies the place of speech."[82] Yet when he first hears musical sound, a playful pantomime ensues:

> *He hears the flute of Felix, stands amazed and pleased, looks around him, snatches at the empty air, and with clenched hands puts them to each*

ear—appears vexed at his disappointment in not possessing the sound;
rushes forward afterwards, again listens, and, delighted with the sound,
steals off, catching at it with his hands.[83]

By many accounts, the audience was won over immediately. A number of reviewers declared this the best scene of the entire work: "[Cooke's] development of first impressions, and natural perceptions, is given with a fidelity to nature truly admirable . . . nothing can be finer."[84] Some went so far as to complain at the monster being presented in such a favorable light when he should simply be the agent of Frankenstein's downfall.[85] The critic of the *Theatrical Examiner* was so charmed by the monster's behavior that he asked, "Why . . . is not Frankenstein made to have some sympathy with the very moral resentment of the being he has animated into a miserable existence?"[86]

In act 2, scene 3, the audience is further drawn to the monster when a sympathetic exchange takes place with the blind old man, De Lacey. Here, music becomes the medium through which relationships are established (it is one of the few scenes in the entire work in which music indicates anything other than the monster's presence): "*De Lacey . . . plays several chords. The Demon enters, attracted by the lute [harp], suddenly perceives De Lacey, and approaches towards him.*"[87] The monster hides when the old man's children enter, but observes their behavior: "*Music.—Felix takes up a hatchet and chops a log of wood. Music.—[Agatha] takes flowers from the basket— Felix is busied cutting the wood.*" The monster imitates them when they leave: "*Music.—De Lacey feels for the Basket which contained [the violets]— the Demon appears to comprehend his wish, and rushes off . . . Music.—The Demon re-enters cautiously and trembling with a handful of flowers, which he gently places in the basket.*" The lute (or harp) communicates to the monster the kindness of De Lacey; the orchestra becomes the agent by which the monster internalizes and imitates such behavior. Although we have no score, we can imagine the generic lyrical, major-mode music that would have accompanied the tableau.

In the eighteenth century, stringed instruments in particular (whether plucked or bowed) were regarded as having therapeutic value for the physical and mental illnesses caused by disordered nerves.[88] As noted above, the aeolian harp had emerged by the end of the century as a model for conceptualizing the body: its vibrations generated thought, feeling, and imagination, bridging the material and spiritual realms. Dedicatory poems and appreciative descriptions in novels as well as scientific analyses of acoustics evoked the harp's magical effect on mind and body on one hand, and offered it as an explanation for their workings on the other.[89] Furthermore, the harp's production of musical vibrations provided a model for the poet's creation of sounds and rhythms, and the poet's reception of vibrations in turn offered a template for that of his ideal reader.[90] By this two-way process, then, the aeolian harp stood for the delight of unmediated communication. In Peake's play, it captures the sympathetic exchange established between the monster and De Lacey, realized through De Lacey's instrument—an idea to which we shall return.

In Milner's play, much less time is spent in demonstrating the monster's sympathetic impulses. Nevertheless, when his destructive urges come to the surface, as Moody has observed, the iconographic legacy of the wild man—a tendency to switch abruptly between demonic violence and gentle tenderness, and a susceptibility to music—helps to soften the monster. When Frankenstein rather belatedly "assumes an attitude of entreaty," we see the monster's frustration: "*[he] expresses that he would willingly have served Frankenstein and befriended him, but that all his overtures were repelled with scorn and abhorrence—then, with malignant exultation seizes on the Child, and whirls it aloft, as if about to dash it down the rock.*"[91] Such scenes of desperate gestural urgency—and there are a number in both plays—have strong roots in traditions of pantomime and melodrama, in which wordless gestures or tableaux were employed routinely at moments of climax to express passions more directly than was possible with speech.[92] The monster's muteness in such scenes engages the audience's sympathy, both highlighting his capacity for feeling pain, and demonstrating his powerlessness.

It follows from this that, in the politicized atmosphere of the illegitimate theaters, the monster's attempted rebellion against Milner's arrogant Frankenstein and his aristocratic patron would have struck another chord with lower-rank audiences and contributed to their sympathy for him. In any case, resolution of the crisis is precipitated by the monster's charmingly depicted sensitivity to music:

> [Emmeline] *pulls from under her dress a small flageolet, and begins to play an air—its effect on the* Monster *is instantaneous—he is at once astonished and delighted—he places the* Child *on the ground—his feelings become more powerfully affected by the music, and his attention absorbed by it—the* Child *escapes to its father—*Emmeline *continues to play, and* Frankenstein *intently to watch its effect on the* Monster. *As the air proceeds his feelings become more powerfully excited—he is*

moved to tears: afterwards, on the music assuming a lively character,
he is worked up to a paroxysm of delight—and on its again becoming
mournful, is quite subdued, till he lays down exhausted at the foot of the
rock to which Emmeline *is attached.*[93]

The effects of the flageolet resonate strongly not only with the "wild man" tradition of the minor theaters, but also with late eighteenth-century descriptions of music's powers. In his *General History of Music* (1789), Charles Burney expounds at length on how music has long been understood as having power over humans, softening the manners, governing the passions, and healing diseases: "the action of musical sounds upon the fibres of the brain" can affect both the mind and the nervous system.[94] In terms of aesthetic pleasure, Burney observes, "It is not the most refined and uncommon melody, sung in the most exquisite manner . . . which has the greatest power over the passions of the multitude . . . the most simple music . . . will be more likely to rouse and transport them." The impact of simple melody on even the least cultivated ears was amply demonstrated in the story of the seventeenth-century composer Stradella, whose murder was thwarted when his bandit assassins heard his beautiful music—a story that appeared, for example, in Richard Eastcott's *Sketches of the Origin, Progress and Effects of Music* (London, 1794) and continued to circulate widely in the early nineteenth century.[95] David Hartley claimed that the appreciation of music derived not only from sensation, but also from association.[96] This principle was explored in more detail by Archibald Alison in his *Essays on the Nature and Principles of Taste* (1790). In the opening essay, Alison locates aesthetic perception at the point when received sensations create trains of thought.[97] In the next essay, he explains the mechanism of association, whereby music acquires an imaginary emotional correspondence or aesthetic signification: the sound of a storm is sublime in that the mind associates it with power and danger (assuming one has experienced a storm before); the same applies to music heard in particular situations. Associations may derive from the equation of emotional states with tempi, textures, and pitches. As Leslie Blasius has noted, although the German romantics disposed quickly of the epistemology of sensation and association in the early decades of the nineteenth century, English discourse on music preserved it—and it is seen in action in these *Frankenstein* plays.[98]

In addition to its physical and associative effects on the individual, music's capacity to foster sympathy between people—and to cultivate reciprocal behavior—was a popular topic in the press, and was often underscored with references to electricity's similarly mysterious powers. A review of Domenico Cimarosa's *Gli Orazi e i Curiazi* (1796) at the King's Theatre in 1814 is typical: at the conclusion of the opera, the "powers" of the tenor Diomiro Tramezzani "were aroused by the sympathy which [Angelica Catalani] produced, and the burst of passion between them was sustained with an emphasis that was felt like electricity in every nerve of the spectators."[99] Likewise, in the scenes from the two *Frankenstein* plays, we have examples of the monster responding instinctively to the sound of music—its vibrations—but also learning to listen and behave through association with emotional states, in so doing establishing an understanding with those around him.

MUSIC AS STIMULANT

In spite of the reserves of sympathy established between the monster and the audience in the early stages of the story, however, frustration gets the better of him: he seeks revenge in a sequence of extended episodes in which music seems to inflame his passions further, and to encourage his violent actions. In Peake's play, when the monster rescues Agatha after she falls into a river (albeit through shock at seeing him), his actions are accompanied by the orchestra: "*Music.—The Demon places Agatha, insensible, on a bench near De Lacey . . . [He] tenderly guides the hand of De Lacey and places it on Agatha.*"[100] This incursion of soft music seems to portray the monster's feelings of kindness and sympathy toward Agatha and the old man—emotions that he had internalized through music in the previous scene. However, when Felix and Frankenstein find the "demon," they shoot at him, whereupon he "*writhes under the wound*," and then: "*in desperation [he] pulls a burning branch from the fire—rushes at them—beholds Frankenstein—in agony of feeling dashes through the portico.*" Peake's own sympathy for the monster is evident in the way this latter is made to act out of desperation and self-preservation. However, during the act finale, desire for revenge takes over: in an extended tableau, the monster is seen to climb up the outside of the cottage, burst through the thatch and—hanging from the rafters—set fire to the building "with malignant joy." He is accompanied by a chorus of onlookers, whose near-gleeful chanting seems to stoke his feelings for revenge: "The fiend of Sin / With ghastly grin! / Behold the Cottage firing!" In the final act, the monster continues on the path of vengeance, carrying off Frankenstein's young brother William (we later learn he is dead), and then murdering Agatha in a scene of musically accompanied pantomime during which Frankenstein—unaware of the monster's presence—speaks aloud his fears for Agatha. Again, convention determined that such scenes were accompanied by the orchestra: music would have established the mood and pacing of the drama, adding fluidity to (and making narrative sense of) the sequence of individual actions, and building tension, effectively goading the monster. We finally hear a "piercing scream": in a large mirror "Agatha appears on her knees with a veil over her head.—The Demon with his hand on her throat— she falls—the Demon disappears."[101]

Though we do not have a score for this drama, we may speculate about how the music sounded on the basis of other works on comparable themes, written in similar theatrical contexts. We know, for example, that the English Opera House boasted a sizeable orchestra. Music for the *Skeleton Lover* of 1830 includes orchestral parts for at least eighteen instruments, including trombone and timpani, and atmospheric passages accompany sections of stage action (typically fragments of up to thirty bars, or in the act finales lengthier passages interspersed with speech). Tremolo strings (as one would expect in a play dealing with the supernatural) feature alongside scalic or chromatic motifs, sustained chords, approaching "footsteps" suggested with rising pizzicato scales, dramatic crescendos and diminuendos, and many *agitato* sections.[102] The monster's music during the violent climactic scenes of Peake's *Frankenstein* would have been similarly graphic, suggesting the brutality of actions that for reasons of taste take place off stage or—as with Agatha's murder—are partially obscured. We might further speculate that the audience would have found this aural and visual spectacle thrilling, enjoying its extravagance.

Music's capacities in these *Frankenstein* dramas to excite as well as to soothe—its dangerous as well as its therapeutic effects—can be mapped onto the evolving model of the aeolian harp: the chaotic force of the wind threatens to overwhelm the capacity of the strings to harmonize. For Coleridge, for example, the self no longer vibrates in sympathy with the universe: "the dull sobbing draft, that moans and rakes / Upon the strings of this Æolian lute" presages "Reality's dark dream."[103] The orchestra—like the harp—consists of both strings and wind, and so in this context has the capacity to "overwhelm" and derange as well as to animate, to calm and to generate sympathy. James Kennaway has explained how the first decades of the nineteenth century saw a shift toward a pathological understanding of music's effects, whereby it was acknowledged as a dangerous stimulant.[104] As noted above, while nerves came to be understood in the context of galvanism and animal electricity, music in turn came to be seen in these terms, as offering a quasi-electrical charge. The apparent overstimulation of the monster's nerves by the chorus and then the orchestra becomes the cause of his violent behavior, of the translation from thoughts of vengeance to destructive acts.

Both dramas conclude with the rapid annihilation of the monster and his creator. In Peake's play an avalanche triggered by Frankenstein's musket shot buries them both. Three years later, influenced by a French version of the story on the Parisian stage, the mountain was replaced with a seascape: the monster swims after and boards Frankenstein's boat: they both die— Frankenstein of despair and the monster of exhaustion.[105] In Milner's play, the monster jumps into the crater of an erupting Etna after stabbing Franken-stein to death.[106] The audiences of both works, having been encouraged to sympathize with a monster whose emerging sense of self and generosity of spirit and whose horrible but clearly motivated acts of revenge have been so effectively communicated through music and mime, were denied a moral resolution. The critic for the *Theatrical Examiner* concluded that Peake's play instead offered a critique of modern society:

> The formation of distorted humanity is an affair of every day recurrence; so much so, indeed, that we were half disposed, on Monday night, to regard this drama as a satire on our Irish system, which creates mon-sters exactly like the over-curious *Frankenstein*, and in the same manner runs about shooting them for being precisely what they have been made. . . . The dramatic monster too was willing to work hard, to cut wood and bear heavy burdens; but the system stood in his way. His kindness was repulsed, its unavoidable prejudices treated roughly, and, in revenge, he sets fire to a cottage! The disguise is too shallow; it is certainly a satire![107]

Bearing in mind the political significance of pantomime at the illegiti-mate theaters alluded to above, it seems likely that at least some members of the audiences at these plays—like the critic of the *Examiner*—would have been disposed to sympathize with the monster's silent rebellion against au-thority. And the monsters' muteness created further resonances: by present-ing music as a palpable expression of vitality, Peake and Milner (and their collaborators) offered their spectators the opportunity to participate in the monster's discovery of the sensory world and even to sympathize with his destructive behavior, as the orchestra acted on the spectators' own nervous systems too. The preface to Mary Shelley's novel suggested that vitalism "af-fords a point of view to the imagination for the delineating of human pas-sions more comprehensive and commanding than any which the ordinary re-lations of existing events can yield."[108] The stage adaptations of *Frankenstein* performed this imaginative act, and in so doing demonstrated the suggestive potential of the relation between music and electricity. Not only was this allusive energy part of the physical, material world and an invisible, transcendent means of communication, but it also mediated between the worlds of science, art and politics.

NOTES

1. Richard Brinsley Peake, *Presumption; or, The Fate of Frankenstein, a Romantic Drama in Three Acts*, in *Seven Gothic Dramas, 1789–1825*, ed. Jeffrey Cox (Athens: Ohio University Press, 1992), 398–99. I use the term "monster" for what is termed (variously) demon, creature, nondescript, and monster in different sources and contexts.

2. *London Magazine*, September 1823, 322–23; cited in Jane Moody, *Illegitimate Theatre in London, 1770–1840* (Cambridge: Cambridge University Press, 2000), 94.

3. The *Theatrical Observer* quoted a placard warning "Do not go to the Lyceum to see the monstrous Drama, founded on the improper work called "Frankenstein" (August 9, 1823); cited in Moody, *Illegitimate Theatre in London*, 194. Various papers, including the *London Morning Post*, noted that it "met with some opposition" at the close (July 29, 1823).

4. *Theatrical Observer*, August 1, 1823.

5. *London Morning Post*, July 30, 1823.

6. *Morning Chronicle* (London), July 31, 1823.

7. Steven Forry identifies eight *Frankenstein* plays on the London stage between 1823 and 1828: Richard Brinsely Peake, *Presumption; or The Fate of Frankenstein* (English Opera House [Lyceum Theatre], July 28, 1823), Henry M. Milner, *Frankenstein; or The Demon of Switzerland* (Royal Coburg Theatre, August 18, 1823), *Humgumption; or Dr Frankenstein and the Hobgoblin of Hoxton* (New Surrey Theatre, September 1, 1823), *Presumption and the Blue Demon* (Davis's Royal Amphitheatre, September 1, 1823), Richard Brinsley Peake, *Another Piece of Presumption* (Adelphi Theatre, October 20, 1823), *Frank-in-Steam; or The Modern Promise to Pay* (Olympic Theatre, December 13, 1824), Henry M. Milner, *Frankenstein; or The Man and the Monster* [in some sources *The Man and the Monster; or The Fate of Frankenstein!*] (Royal Coburg Theatre, July 3, 1826), and John Kerr, *The Monster and Magician; or, The Fate of Frankenstein* (New Royal West London Theatre, October 9, 1826). See Steven Earl Forry, "Dramatizations of Frankenstein, 1821–1986: A Comprehensive List" English Language Notes 25:2 (December 1987), http://knarf.english.upenn .edu/Articles/forry2.html. See also Forry's "The Hideous Progenies of Richard Brinsley Peake: *Frankenstein* on the Stage, 1823 to 1826," *Theatre Research International* 11 (1985), 13–31; and *Hideous Progenies: Dramatizations of* Frankenstein *from the Nineteenth Century to the Present* (Philadelphia: University of Pennsylvania Press, 1990). In addition, Jane Moody mentions *Dr Frankenstein and his Son* (Surrey, 1823); *Illegitimate Theatre in London*, 94. A further seven *Frankenstein* plays were performed between 1821 and 1826 in Paris: it seems that there was exchange and influence across the channel in some of the productions, although the most significant scientific context for the Parisian works was evolution rather than vitality; see my "Monsters and the Mob: Grotesque on the Parisian Stage, 1826–1836," in *Textual Intersections: Literature, History and the Arts in Nineteenth-Century Europe*, ed. Rachael Langford (Amsterdam: Rodopi, 2009), 29–40.

8. This generic indeterminacy was typical of the "illegitimate" aesthetic—the work is referred to variously as "romantic drama," "romance," "melodrama" and "melodramatic opera." In this chapter I tend to use the more neutral "play."

9. Since the eighteenth century, licensing laws had restricted the performance of spoken drama to the patent theaters, Covent Garden and Drury Lane (and the Haymarket for the summer months); other theaters (including the Lyceum) were permitted to stage only burlettas and other musically accompanied light entertainments. In practice, however, the boundaries were not clearly defined, and as discussed below, the situation had become quite confused by the mid-1820s. See Jane Moody, "The Theatrical Revolution, 1776–1843," in *The Cambridge History of British Theatre*, ed. Joseph Donohue (Cambridge: Cambridge University Press, 2008), 199–216.

10. John Hunter had developed an understanding of the body as a self-communicating organism with automatic powers of healing and a force of vitality in the blood: the dispute was between vitalist and materialist interpretations of his work. Simon Schaffer and Steven Shapin, *Leviathan and the Air Pump* (Princeton, NJ: Princeton University Press, 1985), 7.

11. A detailed account of the debate is provided in Sharon Ruston, *Shelley and Vitality* (Basingstoke, UK: Palgrave MacMillan, 2005), 24–73. See also Alan Richardson, *British Romanticism and the Science of the Mind* (Cambridge: Cambridge University Press, 2001), 24–27.

12. John Abernethy, *An Enquiry into the Probability and the Rationality of Mr. Hunter's Theory of Life* [including his first lecture] (London: Longman, 1814), 38; cited in Ruston, *Shelley and Vitalism*, 43. He claimed to be developing the "theory of life" of his former teacher, John Hunter. Abernethy's subsequent publications in the debate included his *Physiological Lectures, Exhibiting a General View of Mr. Hunter's Physiology, and of his researches in Comparative Anatomy* (London: Longman, 1817), and *The Hunterian Oration for the Year 1819* (London: Longman, 1819).

13. William Lawrence, *Introduction to Comparative Anatomy and Physiology* (London: Callow, 1816), 120–21; cited in Iwan Rhys Morus, *Shocking Bodies: Life, Death & Electricity in Victorian England* (Stroud, UK: The History Press, 2011), 43. Lawrence's views were also set out in his *Lectures on Physiology, Zoology, and the Natural History of Man* (London: Callow, 1819), which contained material dating back to 1814.

14. Lawrence, *Lectures on Physiology, Zoology, and the Natural History of Man*, Lecture 3 (1817), 84; cited in Marilyn Butler, introduction to Mary Shelley, *Frankenstein, or The Modern Prometheus: The 1818 Text* (Oxford: Oxford University Press, 1994), xx.

15. Melinda Cooper, "Monstrous Progeny: The Teratological Tradition in Science and Literature," in *Frankenstein's Science Experimentation and Discovery in Romantic Culture, 1780–1830*, ed. Christa Knellwolf and Jane Goodall (Aldershot, UK: Ashgate, 2008), 88. Geoffroy had established the science of teratology (the study of monsters or birth defects) in France, and it was quickly imported to Britain by medical students who had studied with French physiologists.

16. On the politically subversive reputation of chemistry in the wake of the French Revolution, see, e.g., Jan Golinski, *Science as Public Culture: Chemistry and Enlightenment in Britain, 1760–1820* (Cambridge: Cambridge University Press, 1992).

17. Ibid., 53.

18. Morus, *Shocking Bodies*, 45.

19. Lawrence was persuaded to write a letter to the governors of Bridewell and Beth-

lem hospitals, who had voted to suspend him; according to the *Monthly Magazine* (1822) he retracted his "infidel opinions" and promised to "suppress and prevent the circulation of his [*Lectures on Physiology*]"; Ruston, *Shelley and Vitality*, 65.

20. Butler, introduction to *Frankenstein*, xx–xxi. Butler suggests that Shelley employed little more of William Lawrence's work than his critique of Abernethy in the first volume, but volumes 2 and 3 are more wide ranging. Although it aroused only suspicion in 1818, when it was published anonymously, after 1820—when Lawrence's most important work was taken over by radical publishers—the scientific dimensions of the novel must have become more readily decipherable (xxxii–xxxiii).

21. Richardson, *British Romanticism*, 1–38.

22. David Hartley, *Observations on Man; His Frame, his Duty and his Expectation* (London, 1749); see Richardson, *British Romanticism*, 9–10.

23. These ideas are examined by Kimiyo Ogawa, "'Suspended' Sense in *Alastor*: Shelley's Musical Trope and Eighteenth-Century Musical Discourse," in *The Figure of Music in Nineteenth-Century British Poetry*, ed. Phyllis Welliver (Aldershot, UK: Ashgate, 2005), 50–69, esp. 65.

24. Richardson, *British Romanticism*, 6.

25. Shelley Trower, *Senses of Vibration: A History of the Pleasure and Pain of Sound* (New York: Continuum, 2012), 13, 15. Trower is critical of Jonathan Crary's claim that there was a sharp break around 1810 from classical models of vision to a "subjective vision," with a new understanding of sensations as originating in the body rather than the external world, and argues instead for a more nuanced understanding of sensitivity. She offers Coleridge's 1795 poem as an exemplar of this understanding: "And what if all of animated nature / Be but organic Harps diversely framed, / That tremble into thought"; Samuel Taylor Coleridge, "The Aeolian Harp" (1795), in *Samuel Taylor Coleridge: The Major Works*, ed. H. J. Jackson (Oxford: Oxford University Press, 1985), 27–29. Coleridge—and other writers—used the words "harp" and "lyre" interchangeably at this time.

26. Discussed in Erasmus Darwin's *Zoonomia* (1794–1796); see Richardson, *British Romanticism*, 12–16.

27. Richardson, *British Romanticism*, 13. Darwin's ideas overlapped with those of Cabanis, for whom "sensibility" was a physical process that radiated from the brain (18).

28. Charles Bell, *Idea of a New Anatomy of the Brain* (London: Dawsons of Pall Mall, 1811); see Richardson, *British Romanticism*, 30–34.

29. Richardson, *British Romanticism*, 13. As Richardson explains, in the 1790s the connection between unorthodox (French-inspired) science and political radicalism ensured that his views were seen as politically subversive. For Schaffer and Shapin, however, by the end of the eighteenth century, "vitalism had become identified with a controversial and fully fledged materialism, whose political, theological and scientific implications in distancing God from nature, and insisting on the autonomous powers of matter, were fully recognized by both proponents and critics" (*Leviathan and the Air Pump*, 78).

30. This account of the experiment comes from Morus, *Shocking Bodies*, 23.

31. Mary A. B. Brazier, "The Evolution of Concepts relating to the Electrical Activity of the Nervous System, 1600–1800," in *The History and Philosophy of Knowledge of the Brain and its Functions*, ed. F. N. L. Poynter (Oxford: Blackwell, 1958), 215; cited in Richardson, *British Romanticism*, 7.

32. A fellow of the Royal Society; cited in Richardson, *British Romanticism*, 24–25. Although back in 1776, William Cullen had written about restoring the vital principle to

the (apparently) dead by "warming the body, blowing smoke into the lungs, opening veins, etc." See Schaffer and Shapin, *Leviathan and the Air Pump*, 19.

33. Schaffer and Shapin, *Leviathan and the Air Pump*, 25.

34. "Galvanism," *Philosophical Magazine*, 14 (1802), 364–68; cited in Iwan Rhys Morus, *Frankenstein's Children: Electricity, Exhibition, and Experiment in Early-Nineteenth-Century London* (Princeton, NJ: Princeton University Press, 1998), 127. See also a slightly differently worded account in John [Giovanni] Aldini, *An Account of the Late Improvements in Galvanism* (London: Cuthell and Martin, 1803); cited in Richard Holmes, *The Age of Wonder: How the Romantic Generation Discovered the Beauty and Terror of Science* (London: HarperCollins, 2008), 317.

35. As reported in the *Quarterly Journal of Science*, 6 (1818), 283–94; cited in Holmes, *The Age of Wonder*, 128.

36. Morus, *Shocking Bodies*, 27.

37. Charles Kite, "An Essay on the Recovery of the Apparently Dead: Being the Essay to which the Humane Society's Medal was adjudged" (London: Dilly, 1788), *Annual Report* (London: Humane Society, 1788), 225–44; "A (Not So) Brief History of Electrocardiography," *ECG Library*, accessed February 11, 2014, http://www.ecglibrary.com/ecghist.html.

38. Robert Darnton, *Mesmerism and the End of the Enlightenment in France* (Cambridge, MA: Harvard University Press, 1968), 4. Although Mesmer was based in Vienna and Paris, his ideas spread rapidly throughout Europe; he tried—unsuccessfully—to gain official approval for his doctrines from the Royal Academy of Sciences in London.

39. *Examiner*, October 28, 1821.

40. *London Morning Post*, August 19, 1823; the correspondent signed himself as W. Wright.

41. Franz Anton Mesmer, *Mesmerism, A Translation of the Original Scientific and Medical Writings of F. A. Mesmer*, trans. George Bloch (Los Altos, CA: Kauffmann, 1980), 19; cited in Trower, *Senses of Vibration*, 25.

42. James Kennaway, "Musical Hypnosis: Sound and Selfhood from Mesmerism to Brainwashing," *Social History of Medicine* 25, no. 2 (October 2011): 273.

43. Mesmerism had attracted interest in London in the 1790s, but it was not until one of Mesmer's disciples, Richard Chenevix, demonstrated its potential for medical use in 1829 that it gained a foothold. Chenevix published a series of articles in the *London Medical and Physiological Journal* and gave demonstrations at St Thomas's Hospital. English translations of Mesmer followed in the 1840s. See Fred Kaplan, *Dickens and Mesmerism: The Hidden Springs of Fiction* (Princeton, NJ: Princeton University Press, 1975), 14–15.

44. *Literary Panorama and National Register* (1817), vol. 6, cols. 268–69: Charles Taylor reported briefly on activities and publications concerning "animal magnetism" in Germany, Italy, and Russia, and noted that attempts had been made in Britain "to obtain . . . a public establishment, and a professor's chair."

45. Angelo Colò, *Podromo sull'azzione salutare del magnetismo animale e della musica* (Bologna: Giuseppe Lucchesini, 1815), 87. This story is described (with translated excerpts) and examined in detail in Ellen Lockhart, "Giuditta Pasta and the History of Musical Electrification," paper presented at the Annual Meeting of the American Musicological Society, San Francisco, November 10–13, 2011.

46. Lockhart, "Giuditta Pasta and the History of Musical Electrification"; see also Céline Frigau-Manning, "Singer-Machines: Describing Italian Singers, 1800–1850," *Opera Quarterly* 28, no. 3–4 (2012): 230–58.

47. Stendhal, *Vie de Rossini* (Paris: Boulland, 1824), trans. By Richard Coe as *Life of Rossini* (London, 1956), 377; cited in Lockhart, "Giuditta Pasta and the History of Musical Electrification."

48. Extract from Stendhal, *Life of Rossini*, as cited in A. F. Crell, *The Family Oracle of Health* (London: J. Walker, 1824), 1:228, and described in James Kennaway, "From Sensibility to Pathology: The Origins of the Idea of Nervous Music around 1800," *Journal of the History of Medicine and Allied Sciences* 65, no. 3 (2010): 420.

49. See his *Hygëia* (1802); discussed in Kennaway, "From Sensibility to Pathology," 416.

50. James Kennaway, *Bad Vibrations: the History of the Idea of Music as a Cause of Disease* (Aldershot, UK: Ashgate, 2012), 36–37.

51. This is Richardson's summary of Edward S. Reed's position in *From Soul to Mind: The Emergence of Psychology, from Erasmus Darwin to William James* (New Haven, CT: Yale University Press, 1997); see Richardson, *British Romanticism*, 37.

52. For more information on the "Bracknell circle" and Lawrence's relations with the Shelleys (drawn largely from Godwin's diary entries and from correspondence), see Ruston, *Shelley and Vitality*, 86–95.

53. Ibid. See also Butler, introduction to *Frankenstein*, xv–xxi.

54. Richard Holmes has summarized the earlier chronology of the novel as follows: Mary's first ideas date back to 1812, when she heard Davy's public lectures on chemistry; journal entries show that she and Percy were discussing ideas of creating artificial life in 1814, when they eloped to France and Switzerland; see *The Age of Wonder*, 325–26, 331.

55. Ibid., 327.

56. Ibid., 328; Butler, introduction to *Frankenstein*, xxix.

57. Shelley, preface to *Frankenstein*, 3. This excerpt was also published in the *London Morning Post* on July 28, 1823, in anticipation of the first performance of Peake's play, and on various playbills.

58. Shelley, "Appendix A: "Author's Introduction to the Standard Novels Edition (1831)," in *Frankenstein*, 195–96.

59. In Peake's play, Frankenstein acquires a (comic) assistant, Fritz, and there are several love affairs in train, but the story is recognizably that of Shelley; Milner's play departs further—shifting the location to Sicily and characterizing Frankenstein as a rather arrogant celebrity scientist (with abandoned lovers in his wake) who has a room in the pavilion of Prince del Piombino in which to carry out his experiment.

60. See Jane Moody, "The Theatrical Revolution."

61. Letter to Leigh Hunt (*Letters*, 1:378); cited by Stephen C. Behrendt, "Presumption; or The Fate of Frankenstein," August 2001, *Romantic Circles*, accessed February 11, 2014, http://www.rc.umd.edu/editions/peake/apparatus/drama.html.

62. Elaine Hadley, *Melodramatic Tactics: Theatricalized Dissent in the English Marketplace, 1800-1885* (Stanford, CA: Stanford University Press, 1995), 54.

63. Hadley describes the responses to the disturbance, which centered on the hefty rise in ticket prices following the rebuilding of the theater after a fire. At the end of the performance, the audience refused to leave, and the manager sent for the police; ibid., 55–56.

64. Moody, *Illegitimate Theatre in London*, 95.

65. See, e.g., Franco Moretti, "The Dialectic of Fear," *New Left Review* 136 (1982): 67–85; Chris Baldick, *In Frankenstein's Shadow: Myth, Monstrosity, and Nineteenth-Century Writing* (Oxford: Clarendon Press, 1987).

66. This is not to suggest that the *Frankenstein* audiences were all from the lower classes, of course (see below). Jim Davis has cautioned against generalizing about audiences or overcategorizing by class and offers a detailed assessment of the factors in his study of the Surrey and Coburg theaters (with Victor Emeljanow), "New Views of Cheap Theatres: Reconstructing the Nineteenth-Century Theatre Audience," *Theatre Survey* 39, no. 2 (1998): 53–72.

67. Cox's edition is based on the two extant performing texts: a Larpent copy (LA 2359), where it is called "*Frankenstein*, A Melo-Dramatic Opera in 3 Acts," and its publication in Dick's Standard Plays, number 431 (London, 1865?), which seems to be based on a manuscript held at the Bodleian (M. Adds. 111.2.11); see Jeffrey Cox, "Introduction to *Presumption; or the Fate of Frankenstein*," in *Seven Gothic Dramas*, 385–87. Forry's website (see note 7) includes Cox's edition together with images, reviews, and a bibliography.

68. *London Morning Post*, July 29, 1823.

69. The text for the 1823 version does not appear to have survived; I have used the 1826 version: H. M. Milner, *Frankenstein; or The Man and the Monster!, a Peculiar Romantic, Melo-Dramatic Pantomimic Spectacle in Two Acts* (London: Duncombe, n.d.), also available at http://knarf.english.upenn.edu/Milner/milner.html.

70. Information about the play and its source is given in Cox, *Seven Gothic Dramas*, 385.

71. Simon During, "'The Temple Lives': The Lyceum and Romantic Show Business," in *Romantic Metropolis: The Urban Scene of British Culture, 1780-1840*, ed. James Chandler and Kevin Gilmartin (Cambridge: Cambridge University Press, 2005), 204–24. The Lyceum had been built in the 1770s as an exhibition and concert hall. It was licensed for musical works in 1809 and was rebuilt in 1816 as the English Opera House, acquiring a license to stage musical farces and ballad operas during the summer; it burnt down in 1830. See Cyril Ehrlich, Simon McVeigh, Michael Musgrave, "London (i), §VI; Musical life: 1800–1945, 1: The stage, (i) Opera," *Grove Music Online* (Oxford University Press), accessed February 11, 2014; http://www.oxfordmusiconline.com.

72. Davis and Emeljanow, "New Views of Cheap Theatres." The Coburg (renamed the Royal Victoria in 1833) had been built as a speculative venture to challenge the Surrey Theatre in the wake of the opening of Waterloo Bridge in 1817.

73. See Louis James, "Frankenstein's Monster in Two traditions," *Frankenstein, Creation and Monstrosity*, ed. Stephen Bann (London: Reaktion Books, 1994), 77–94. In Charles Dibdin's *The Wild Man* (Sadler's Wells, 1809, in which O. Smith later starred), for example, the wild man is affected by Artuff's flute-playing: "[the Wild Man] appears quite softened by the melody—which Artuff increases to 'moderato'—the eyes of the Wild Man brighten, and he expresses joy—Artuff increases to 'furioso':—this strain excites the Wild Man's feelings to passion and ferocity . . . Artuff plays 'affetuoso';—this softens the Wild Man, who cries" (87). On the eloquence of the "wild man," see also Moody, *Illegitimate Theatre in London*, 94.

74. See costume descriptions in Milner, *Frankenstein*, 3; this edition includes as a frontispiece an engraving of "Mr. O. Smith as the Monster in Frankenstein" that seems to tally with this (though his legs are not visible)—he also sports a large curly black wig.

75. *London Morning Post*, July 30, 1823. When Smith took over temporarily from Cooke in Peake's play in 1827, several reviewers compared their talents—and refused to choose between them; see, for example, *London Morning Post* (10 July 1827).

76. Shelley, *Frankenstein, or The Modern Prometheus: The 1818 Text*, 81.

77. Ibid., 85, 94.

78. Milner, *Frankenstein*, 11 (act 1, scene 3).

79. Peake, *Presumption*, 397–8 (act 1, scene 3).

80. Chandler, *An Archaeology of Sympathy*, 243; as Chandler mentions (246), the monster declares that "If I have no ties and no affections, hatred and vice must be my portion."

81. Ibid., 227–28.

82. *Oxberry's Dramatic Biography and Histrionic Anecdotes*, vol. 1 (London: G. Virtue, 1825), cited in Jim Davis, "Presence, Personality and Physicality: Actors and their Repertoires, 1776–1895," in *The Cambridge History of British Theatre*, ed. Joseph Donohue (Cambridge: Cambridge University Press, 1995), 278.

83. Peake, *Presumption*, 405 (act 2, scene 2).

84. *London Morning Post*, July 30, 1832.

85. *Times* (London), July 29, 1830.

86. *Theatrical Examiner*, August 3, 1823.

87. Peake, *Presumption*, 407 (act 2, scene 3). "Harp" and "lute" seem to be interchangeable in the sources: Cox uses "lute" (from Larpent), but notes that Dick's Standard Plays has "harp."

88. Trower, *Senses of Vibration*, 34.

89. Ibid., 9; see also Thomas Hankins and Robert Silverman, "The Aeolian Harp and the Romantic Quest of Nature," in *Instruments and the Imagination* (Princeton, NJ: Princeton University Press, 1995), 86–112.

90. Trower, *Senses of Vibration*, 9.

91. Milner, *Frankenstein*, 24 (act 2, scene 4).

92. Indeed, for some observers the monster's discovery of his senses would have evoked the experience of the statue, whose senses gradually opened onto the world of experience, in Etienne Bonnot de Condillac's *Traité des sensations* (1754), a work that was still current in the 1820s. For more on this, and its influence on "reform" pantomime, see Ellen Lockhart, "Alignment, Absorption, Animation: Pantomime Ballet in the Lombard Illuminismo," *Eighteenth-Century Music* 8, no. 2 (2011), 244–46. Leslie David Blasius has even suggested that "Shelley's creature is Condillac's statue given literary flesh"; see his "The Mechanics of Sensation and the Construction of the Romantic Musical Experience," in *Music Theory in the Age of Romanticism*, ed. Ian Bent (Cambridge: Cambridge University Press, 1996), 5.

93. Milner, *Frankenstein*, 24–25 (act 2, scene 4). The flageolet was the instrument that young girls would take up, and there does not appear to be any greater significance in this choice.

94. Charles Burney, *A General History of Music: from the earliest ages to the present period (1789)*, 2 vols. (London: Foulis, 1935), 1:159. Music's effects on animals are less conclusive, however: birds, "so fond of their own music" are not charmed by ours (though they may compete "to surpass it in loudness"), and dogs and cats "will howl . . . as if the sound were too much for their nerves to bear" (159–60).

95. In his lectures on music history at the Royal Institution and elsewhere in the 1810s, William Crotch categorized music into the beautiful, the sublime and the ornamental, which mapped onto differing levels of refinement of the listener: "the invention of whatever is sublime has been esteemed the greatest effort of the human mind. . . . The lowest and least estimated is the ornamental"; William Crotch, *Substance of Several Courses of Lectures on Music, read in the University of Oxford and in the Metropolis* (London, 1831),

ch. 1–3; cited in Peter le Huray and James Day, *Music and Aesthetics in the Eighteenth and Early-Nineteenth Centuries* (Cambridge: Cambridge University Press, 1988), 248–49.

96. David Hartley (1705–1775) in his *Observations on Man*, a work that was still current at the end of the century.

97. Blasius, "The Mechanics of Sensation and the Construction of the Romantic Musical Experience," 5.

98. Ibid., 17.

99. *Morning Chronicle* (London), May 26, 1814.

100. Peake, *Presumption*, 413 (act 2, scene 5).

101. Ibid., 423 (act 3, scene 3).

102. George Rodwell's score for Edward Fitz Ball's *Skeleton Lover* (1830; performed at the Adelphi, as the Lyceum had burned down earlier that year) are preserved at the British Library (Add MS 33814); I am grateful to James Q. Davies for directing me to this source. The parts are bound into a single volume and comprise leader (violin 1), violin 2 (×2), viola, cello, double bass, flute, oboe, clarinet (×2) bassoon (×2), horn (×2), trumpet, trombone, timpani; additional parts (for a single number in each case) are provided for harp and triangle. The score includes several sung numbers (arias, duets, chorus), some dances, entrance and exit music, very short "punctuating" music (typically four to eight bars) and lengthier passages to accompany stage action—as described above. The forces at the Lyceum for *Frankenstein* are likely to have been greater, thirty-two strong for a performance of Weber's *Sylvania* [*Silvana*] in 1829 according to a reviewer in the *Examiner*; cited (without full reference) in Theodore Fenner, *Opera in London: Views of the Press, 1785–1830* (Carbondale: Southern Illinois University Press, 1994), 574–75.

103. Coleridge, "Dejection: An Ode," (1802) in *Samuel Taylor Coleridge*, 114–18; cited and discussed in Trower, *Senses of Vibration*, 32–34. Historically, stringed instruments were associated with reason and order, wind with the threat of mental or physical derangement.

104. Kennaway, "From Sensibility to Pathology."

105. The new ending is reported in the *London Morning Post*, September 21, 1826.

106. This may have been an obscure reference to the self-igniting Italian mud volcanoes that had fed into theories about electricity at the end of the eighteenth century. Louis James has noted that although the monster's behavior does not secure the audience's sympathy in itself, Frankenstein is presented as a heartless seducer who has abandoned his former lover and their child, and develops the moral implications of his neglect of his family in ways unthinkable in Shelley's novel; James, "Frankenstein's Monster in Two Traditions," 89.

107. *Theatrical Examiner*, August 3, 1823. The reference to the "Irish system" seems to refer to the Irish immigration to England during the first decades of the nineteenth century: a scheme was implemented by the British government in 1823–1825, whereby 2,500 Irish Catholic smallholders were given free passage and land grants. See R. F. Foster, *Modern Ireland, 1600–1972* (London: Penguin, 1990).

108. Shelley, preface to *Frankenstein*, 3.

CHAPTER EIGHT

Engine Noise and Artificial Intelligence: Babbage's London

GAVIN WILLIAMS

In 1914, a week before the guns of August, Baron Moulton of Bank delivered the inaugural address of the John Napier Society, founded in honor of its namesake, the sixteenth-century British inventor of the logarithm.[1] Napier's discovery, Moulton proposed, brought glory to his land, while also conferring more general benefits on humankind by increasing "the powers of the human mind as a practical agent." This he had achieved through speeding up arithmetical calculation: in 1614, exactly three centuries earlier, Napier published a landmark series of numerical tables that demonstrated how the multiplication of large numbers could be reconceived in terms of addition. Mulling the logarithm's significance for the history of science, Moulton wondered what had spurred Napier's insight. As a barrister deeply engaged in intellectual property, Moulton opted for a realist view: during his career at the Bar he had had the opportunity to notice the "circuitous" routes by which inventions were effected; Napier, like other innovators, had "groped in the dark" (for over twenty years) before perfecting his invention.[2] In the end he had emerged victorious—thereby demonstrating "the persistent effort of a great mind to perform a task which it has deliberately set to itself."[3] Everything depended on the eventual completion of that self-imposed mission: without the successful publication of his tables, the logarithm would have remained a scientific nonentity.

To underscore Napier's achievement, Moulton contrasted the fulfillment of an original goal with its opposite: with failure. In particular, he drew his audience's attention to the recent example of Charles Babbage's calculating engines. In the 1820s, Babbage began designing and constructing elaborate machines intended to compute, among other things, logarithmic tables. He was still sketching blueprints in the 1870s, during the final years of his life. It was around this time that Moulton, then an undergraduate math star at

Cambridge University, paid a visit to Babbage at his London residence, where he was given a tour of three domestic workrooms. The different spaces represented phases of the machine's development:

> In the first room I saw the parts of the original Calculating Machine, which had been shown in an incomplete state many years before and had even been put to some use. I asked him about its present form. "I have not yet finished it because in working at it I came on the idea of my Analytical Machine, which would do all that it was capable of doing and much more. Indeed, the idea was so much simpler that it would have taken more work to complete the Calculating Machine than to design and construct the other in its entirety, so I turned my attention to the Analytical Machine." After a few minutes talk we went into the next work-room, where he showed and explained to me the working of the elements of the Analytical Machine. I asked if I could see it. "I have never completed it," he said, "because I hit upon an idea of doing the same thing by a different and far more effective method, and this rendered it useless to proceed on the old lines." Then we went into the third room. There lay scattered bits of mechanism, but I saw no trace of any working machine. Very cautiously I approached the subject, and received the dreaded answer, "It is not constructed yet, but I am working on it, and it will take less time to construct it altogether than it would have taken to complete the Analytical Machine from the stage in which I left it." I took leave of the old man with a heavy heart.[4]

The calculating machine that Moulton remembered had been given another name by its inventor: the difference engine. By the mid-nineteenth century, it had become notorious as one of the earliest examples of public funds squandered in the pursuit of unintelligible science.[5] Beginning in 1823 with a grant of £1,500, Babbage received large subventions from the British Treasury, grants that continued sporadically until 1842—that is, until Babbage expressed his desire to abandon his early project in favor of starting work on his analytical engine. Understandably, the government was reluctant to write off the difference engine, which had already consumed so many funds. After seeking reassurances from the Royal Society, the Chancellor of the Exchequer granted Babbage one last payment—on condition that he strive to realize his original design. He grudgingly accepted the government's terms and applied renewed efforts to his (quietly modified) difference engine no. 2. But this project was similarly destined never to be finished.[6] By the end of the

1840s, with no more money and with a working machine still a distant prospect, Babbage finally gave up. The British press lampooned the harebrained extravagance of an archetypal mad scientist.[7] As the *Morning Chronicle* put it, affecting sympathy with Babbage's frustrations:

> After twenty years' ceaseless labor of an intense description, together with an expenditure of a large fortune in hard cash, the whole undertaking was smashed—Mr. Babbage's hopes frustrated, besides his labor and money thrown away. It would be difficult to imagine a more bitter disappointment, the great work of a celebrated man's life suddenly reduced to a mere bagatelle.[8]

This public scandal lingered in the collective memory, at least until the eve of the First World War, when Moulton could revisit the story expecting his audience would remember the gist. In calling on distant memories, Moulton was passing on a myth—one that he elaborated through his private tour of the workrooms. According to his telling, each door revealed a greater horror: a progressively more ambitious machine aborted at an ever-earlier stage of construction. Like an aged Bluebeard, Babbage was haunted by these phantom love-objects lain to waste around his home: his property on Dorset Street in Marylebone had become a kind of sanctuary in which he could hold onto dreams of realizing his life's work.

Moulton's domestic psychodrama might give us pause, for it bears comparison with another urban legend associated with the Babbage household. It was from here that Babbage launched his campaign against "Street Nuisances"—the name given to his pamphlet-length tirade against organ grinders and itinerant brass band players, whom he targeted as prime social menaces and irritating polluters of the urban environment. As he put it in 1864:

> During the last ten years, the amount of street music has so greatly increased that it has now become a positive nuisance to a very considerable portion of the inhabitants of London. It robs the industrious man of his time; it annoys the musical man by its intolerable badness; it irritates the invalid; deprives the patient, who at great inconvenience has visited London for the best medical advice, of that repose which, under such circumstances, is essential for his recovery, and it destroys the time and the energies of all the intellectual classes of society by its continual interruptions of their pursuits.[9]

At the time of its publication, Babbage's tract became well known within the context of a wider debate over a proposed law controlling street musicians. A key issue was the right to silence of the convalescent patient—or, more obviously, the Victorian male professional.[10] While there appeared to be a certain public-spiritedness about Babbage's complaint, his reference to general medical concerns and the needs of the industrious intellectual are disingenuous, belying his own carefully nurtured pet obsessions. He was much preoccupied by his own failing health; and, as one member of the self-proclaimed intellectual classes of society, he felt his time and energies were being dissipated by music coming from the street. Not content to suffer this assault in silence, during the 1850s and '60s he decided to confront these aggravations, advancing a strategic offensive against the noises of his neighborhood.

Taking my historical coordinates from this contested London topography, I explore the links between Babbage's favored geriatric occupations: between his ever-ongoing work on the difference engine on the one hand and his crusade against street musicians on the other. In his trenchantly anti-autobiographical autobiography (he deemed the genre too sedentary for a busy "philosopher"), Babbage hinted at an intimate connection between the two: "On careful retrospect of the last dozen years of my life, I have arrived at the conclusion that I speak within the limit when I state that one-fourth part of my working power has been destroyed by the nuisance against which I have protested."[11] Babbage's reference to the annihilation of intellectual work points to an unusual perspective on the issue of street music, and can begin to evoke for us his strained posturing vis-à-vis the sonic environment. In the pages that follow, I will try to elucidate his idiosyncratic views, and to use them as a vantage point from which to hear afresh the public and parliamentary debates over controlling London's street music in the 1860s. I hope ultimately to reassess musical activity in the city in the light of broader ideologies of progress—and its retardation—in the urban industrial-scientific economy. The mechanical wreckage at the elderly Babbage's home might serve (pace Moulton) less as a parable of scientific failure than as an allegory of progress's opposite: a negative force that bodied forth as street noise.

MATTERS OF DIFFERENCE

To understand what remained of the difference engine in Babbage's home in the 1860s, first we need to roll back the decades and try to recapture the vigor with which the invention burst on to the scientific scene. In June 1822, Babbage announced his discovery that the "method of differences" (I will return to this term) could enable machinery to calculate astronomical tables; he claimed that a simplified working model had already been built.[12] At roughly the same time, he wrote a long, impassioned letter to Sir Humphry Davy, president of the Royal Society, explaining in detail the rationale behind the invention:

> The intolerable labor and fatiguing monotony of continued repetition of similar arithmetical calculations, first excited the desire, and afterwards suggested the idea, of a machine, which, by the aid of gravity or any other removing power, should become a substitute for one of the lowest operations of the human intellect.[13]

Babbage justified the machine above all as a labor-saving device, one that would also serve to make arithmetical calculation both much faster and entirely free of human error. Further elucidating the possible benefits of substituting machinery for humans, Babbage drew on the example of French mathematician Gaspard de Prony, who, in the wake of the Revolution and at the instigation of its interim government, produced a series of logarithmic and trigonometric tables, with numerical values worked out to unprecedented degrees of precision (to twenty-five decimal places in the case of sine values). This painstaking work—which had occupied Prony and his team of more than eighty human computers for several years—Babbage declared, could have been delegated to a machine such as the one he envisioned.

This was an impressive claim, one designed to stimulate Davy's interest while also (crucially) eliciting the support of the Royal Society—Babbage's machine would require a large financial outlay. Yet by invoking Prony's illustrious tables, Babbage implicitly made a case for a return on the investment.[14] His engine would not only save the British Treasury the £5,000 it intended to offer the French government for a copy of Prony's tables; it would also reduce the number of human beings employed in making any future tables that might be required from eighty to just one or two. Babbage's mechanical revision of Prony's venture, although strikingly original, was characteristic of broader early nineteenth-century views about calculation. As Lorraine Daston has argued, calculation had until recently been a highly regarded faculty of the intellect, but in the years after 1800 was increasingly associated with drudgery and unskilled labor.[15] Whereas Prony's tables were feted in revolutionary France as monuments to mathematical and political progress, Babbage saw their main contribution in industrial terms, as the quotation above suggests. He went on to itemize the thoroughgoing division of labor inside Prony's workshop. In charge were a few master mathematicians (including

Prony himself) who could manipulate algebraic formulae; below them were a team of *calculateurs*, who converted the algebra into numerical values; and at bottom were those with no knowledge of algebra, the ranks of so-called *ouvriers*, who were required to crunch the numbers by means of addition and subtraction only.

Thus, in the early 1820s, Babbage could boast that his machine had the potential to replace a considerable labor force almost entirely. The task of arithmetic calculation could be delegated to cogs and cams, leaving to human effort only the computing of initial values to be fed into the machine. (Babbage decided early on that the difference engine should be capable of printing its results.) His prospective engine drew luster from a contemporary discourse that linked mechanization to industrial progress—a theme picked up in 1823 by the earliest journalistic description of the machine, which compared its probable effects on science with "those rapid improvements in the arts which have followed the introduction of the steam engine."[16] Embedded in this gloss was an emerging analogy between scientific progress and progress in the (mechanical) arts—one that historians of science have detected across a range of phenomena from this period.[17] Specific to Babbage, however, and typically quirky, was the notion that Britain's dynamic factory system should provide a model for efficient scientific practice; and indeed that manufacturing processes could simulate operations that took place within the mathematician's mind.[18] Babbage self-consciously endowed his prospective machines with attributes of cognition, such as memory, recall, and foresight.[19] At their most ambitious, his machines represented hope for a grand three-way synthesis: of the factory system *as* human intelligence *as* calculating machine.

However, this imaginary equation—and with it, the prospect of artificial intelligence—crumbled under the pressure of practical exigencies. As Babbage quickly discovered, he had underestimated the work involved in building the machine: a huge undertaking that would oblige him to sustain a team of engineers and craftsmen throughout the 1820s and '30s. By the time he began work on the analytical engine in 1834, Babbage decided that he needed a dedicated space, and so adapted the land surrounding his residence on Dorset Street.[20] He transformed his home into a mini-factory, complete with workshops and offices—thus intensifying, albeit with the promise of ultimately overcoming, the human industry required for Prony's tables (although, of course, without Prony's numerical results). Yet the productivity of Babbage's domestic manufactory was undermined from the outset. His director of engineering, Joseph Clement, refused to relocate to the new premises, citing disputes over pay and working conditions, bringing work on the machines to a fifteen-month standstill.[21] This bout of small-scale industrial action suggests that the conditions within the workshop were far from ideal; Babbage subsequently confessed that greater progress might have been made had his relations with Clement been better.[22]

There was a more fundamental glitch in Babbage's project that explains why his efforts to displace human intelligence with machines were subject to constant frustration. As Simon Schaffer has argued, Babbage sought always to minimize the human assistance required in the process of calculation, ruling out more flexible interactions between humans and machines.[23] By insisting that the human role should be reduced to data input (a task that would nevertheless have required mathematical training), Babbage attempted what we might call a grandiose concealment of labor—one that could supply the illusion of machine intelligence.[24] Schaffer identified Babbage as the precocious instigator of what would later become a pronounced tendency in computer culture—returning, for example, in Turing's experiments in the mid-twentieth century—that bestows metaphors of cognition on machines, all the while occluding the kinds of human intelligence required to make (and make use of) them. Schaffer rendered newly visible these obscured human-machine interactions, suggesting that the vogue for intricate automata in early nineteenth-century London can be particularly illuminating: in domestic and public displays of automata, the concealment of human skill, and the wonderment it elicited, was once again at issue, encouraging the same transposition of intelligence from craftsman to machine.[25]

And yet, in the exhibition of automata—and of Babbage's calculating machines—a vector also pointed in the opposite direction, linking humans to mechanisms by means of the intelligence involved in attending to them. Significantly, Babbage described this attentive intelligence in aural terms, as the effort involved in the act of listening to the difference engine. Consider, for example, Babbage's explanation to the lay reader of how the difference engine could reconfigure the "mental division of labor." His attempt at simplification by way of analogy—an elaborate thought experiment that shows how a particular mechanism might calculate the series of square numbers— devolved into relentless and baffling tintinnabulations:

> Let the reader imagine three clocks, placed on a table side by side, each having only one hand, and each having a thousand divisions instead of twelve hours marked on the face; and every time a string is pulled, let them strike on a bell the numbers of the divisions to which their hands point. Let him further suppose that two of the clocks, for the sake of distinction called B and C, have some mechanism by which the clock C ad-

vances the hand of the clock B one division, for each stroke it makes on its
own bell; and let the clock B by a similar contrivance advance the hand of
clock A one division, for each stroke it makes on its own bell. With such
an arrangement, having set the hand of clock A to the division I., that of B
to III., and that of C to II., let the reader imagine the repeating parts of the
clocks to be set in motion continually in the following order: viz. pull the
string of clock A; pull the string of clock B; pull the string of clock C. If
now only those divisions struck or pointed at by the clock C [sic; Babbage
meant clock A] be attended to and written down, it will be found that
they produce the series of the squares of the natural numbers.[26]

Here the difference engine is presented as the mechanization of the method
of differences. Clock C adjusted the hand of clock B, which in turn moved
the hand of clock A, while clock C itself never changed from its initial value
of two: it represented constant "difference" of a second order (because two
clocks away from clock A, which showed the result). Babbage proposed that
his difference engine should be able to compute twelve such orders of difference, which would correspond to numerical series much more complex than
the square numbers.[27]

It is worth lingering over Babbage's imaginary machine. In addressing the
lay reader, he conjured up a mechanical fantasy that described the difference
engine as a parlor trick. We are offered strange clocks with miniscule divisions whose hands do not tell the time, but instead point to number series.
The presentation of these series is doubled by the ringing of bells—as though
to prove the results were independent of (abstractable from) the mechanism.
Yet the illusion that the mechanism could *tell* the series of square numbers,
like a grandfather clock tells the hours of the day, relied on a twofold concealment of labor: that of knowing what to listen for in the first place—the
numerate training requisite for setting the initial values—as well as that of
the attentive listening involved in counting the pealing of the bells. Their
imagined sounds enhanced the illusion that the machine itself could perform the calculation, while distracting from the human skills and attentions
that would be expended.

This kind of expert listening became reality when a partial "difference
engine no. 1" was displayed at London's 1862 International Exhibition. For
the occasion, three bells were added to the machine.[28] However, unlike their
imaginary counterpart, these real bells did not toll out all numerical values;
they were rung whenever a particular number series—the computed results,
along with the first and second orders of difference, respectively—plunged
from positive to negative values. The bells thus signaled crucial turning

points (zero values) in the computational process—important moments in-
terpretable only by algebraically informed listeners. Not long after the exhibition, Babbage received a guest, evidently less mathematically adept than
himself, who wanted to experiment with the machine:

> Several weeks after the machine had been placed in my drawing room, a
> friend came by appointment to test its power of calculating Tables. After
> the Engine had computed several Tables, I remarked that it was evidently
> finding the root of a quadratic equation; I therefore set the bells to watch
> it. After some time the proper bell sounded twice, indicating and giving
> the two positive roots to be 28 and 30. The Table thus calculated . . . really
> involved a quadratic equation, although its maker had not previously ob-
> served it.[29]

We are by now familiar with the notion that the difference engine carried out
the active, intelligent work—it "computes," "finds," "watches," "gives" results.
By contrast, Babbage merely registered (remarked on) what the bells were
telling him, while his stupefied friend learned from the machine—Babbage
acting as a passive conduit—that the equation he had devised must have been
a quadratic one.[30] The ruse of machine intelligence was enabled by sounds,
which transposed human skills and operations from a visible to an aural
domain.

STREET MUSIC MACHINES

When the difference engine returned to Babbage's drawing room in 1863, it
entered the midst of a community warzone: his Dorset Street home had become the site for a sustained attack on (and by) street music. By the early
1860s, the battle lines were drawn. On one side, there was Babbage himself
and a few like-minded proprietors of London's professional middle classes.[31]
Babbage received powerful, if intermittent, support from various establishment institutions: the Metropolitan Police Service, whose officers could arrest an offending street musician; the local magistrate, who might—or, as
was often the case, might not—fine said musician up to forty shillings; and,
at a greater remove, nonresident members of Parliament, such as Lord Westmeath and Michael Bass, both of whom (with Babbage's campaign in mind)
presented bills to Parliament against street music in the metropolis.[32] On the
opposite side of the ramparts were Babbage's less music-phobic neighbors,
who not only encouraged street musicians but even sponsored them to play
outside his home, expressly to irritate him. The "rough music treatment"

took place not only during his working day, but also while he slept: one night in December 1863, Babbage was awoken at 1:00 a.m. by a brass band firing up outside his bedroom window. A few weeks later he discovered that these musicians, urged on by his neighbors, had held a midnight rally to coordinate the exact moment of their attack.[33]

More precarious in relation to this middle-class battle of wills were the street musicians themselves: the itinerant brass-band players and organ grinders who toured the metropolis from early morning to late evening, returning at night to accommodations in the poor eastern districts of the city around Holborn Street and Farringdon Road.[34] A significant number of these musicians were not native to Britain, but were migrant workers indentured to compatriot entrepreneurs. Such was the unsteady constituency of Italian organ grinders against whom Babbage railed. Many of them, at least during the early 1860s, were wandering journeymen from the mountainous villages of the Val-di-Taro near Parma.[35] They were poor, often young men and sometimes children who arrived in London unable to speak English and usually without a musical instrument. Instruments could be rented on a daily basis—much like the white mice, monkeys and dancing dogs used by animal exhibitors—from local overlords of the street entertainment trade: available for hire were string-based hurdy-gurdies and the (more common) pipe-based, wind-up barrel organs. The latter ranged in size from the relatively small portable version that could be attached to the body by a shoulder strap, to the huge barrel piano varieties, which paraded aboard large wheelbarrow-like handcarts.[36] Babbage awarded these mobile music machines first place in his checklist of "instruments of torture permitted by the Government to be in daily and nightly use in the streets of London."[37] He estimated (perhaps conservatively) that there were a thousand organ grinders in circulation around the city at any given time.

This migrant industry represented a collective sonic force—one that, according to Babbage, necessitated an opposition cognizant of the economic sources of its cumulative power.[38] In "Street Nuisances," he broke the problem down into its three most significant variables: encouragers, performers, and instruments. Each needed to be tackled on its own terms. Most problematical were the encouragers, the public majority, who were likely to be offended by his tract. He picked off the most vulnerable factions first, gradually working his way up to his real target: frequenters of public houses, servants, children, and prostitutes ("ladies of elastic virtue and cosmopolitan tendencies"; thus far a roll call of the political underclass in mid-Victorian Britain), "visitors from the country," and ultimately "titled ladies; but these are almost invariably of recent elevation, and deficient in that taste which

TABLE 8.1. Babbage's register of street nuisances
showing performing monkeys alongside brass bands
and organ grinders during one week in July 1863.

	Brass bands	Organs	Monkeys
July 3	3	—	—
July 4 (stone hit me)	—	2	—
July 5	—	1	1
July 6	—	1	—
July 7	—	1	—
July 9	—	1	—
July 10 (Tuesday— great mob)	—	1	—

Source: Adapted from Babbage's letter to Sir Michael Bass, published in
Street Music in the Metropolis (London: John Murray, 1864), 20.

their sex usually possess."[39] As this rebarbative flourish makes clear, his opposition to street noise was allied with hatred of women and of the working classes. Babbage's sly insinuation, clearly directed at his nouveau-riche neighbors around Manchester Square, was that street music was feminine, low-bred, and immoral.[40] Meanwhile, he recommended legal and political measures that could be taken against street performers and their borrowed instruments: that the police seize barrel organs (only to be returned on payment of a fine); that the government force magistrates to take a hard line on vagabond musicians; and if less stringent measures should prove futile, that street musicians in London be banned outright. Thus he rounded off his implacable manifesto, with the prospect of expelling a foreign menace: the xenophobic presentiment of a noiseless urban order.

Much like his calculating engines, Babbage thought of London as a perfectible machine, which could be constantly improved by elimination of the human element.[41] With a view to the latter's eventual demise, he began to take detailed observations of the noises circulating within the city, keeping a day-by-day log of the street nuisances as they passed his home. Table 8.1 shows the first seven entries in Babbage's register of disturbances on Dorset Street. He continued his record of brass bands, organ grinders, and monkey exhibitors for another 90 days, during which period he counted 165 separate infractions.[42] Babbage subsequently described the protocol he adopted in the case of incoming intruders:

> Whenever . . . an itinerant musician disturbed me, I immediately sent
> out, or went out myself, to warn him away. At first this was not suc-
> cessful; but after summoning and convicting a few, they found out that
> their precious time was wasted, and most of them deserted the immedi-
> ate neighborhood. This would have succeeded had the offenders been few
> in number; but their name is legion: upwards of a thousand being con-
> stantly in London, besides those on their circuit in the provinces. It was
> not, however, the interest of those who deserted my station to inform
> their countrymen of its barrenness; consequently, the freshly-imported
> had each to gain his own experience at the expense of his own and of my
> time. Perhaps I might have succeeded at last in banishing the Italian nui-
> sance from the neighborhood of my residence; but various other native
> professors of the art of tormenting with discords increased as the license
> of these Italian itinerants was encouraged.[43]

Extraordinary here—and exceptional in the broader attack on street music
at the time—was Babbage's proactive stance. It is as though he alone could
keep itinerant musicians away from Dorset Street and the residential district
surrounding Manchester Square. These musicians were doubly foreign—
intruders both on the nation and, more offensively, on his dignified, well-
to-do locale. Babbage pursued a solo campaign against a much larger system of musical migration (an effort that was doomed to failure, as the above
quotation suggests) and, more generally, against the lower orders of society
as a whole: "I have been compelled individually to resist this tyranny of the
lowest mob, because the Government itself is notoriously afraid to face it."[44]
We might well wonder: how exactly did the venerable philosopher square up
against this anonymous multitude?

A more detailed example of Babbage's tactics can serve to show what happened when he personally went on the offensive. In one carefully described
scenario, Babbage depicted himself suffering the onslaught of invading musicians—this time accompanied by a singing chorus of "shoeless children" and
their "ragged parents"—sent by his neighbors to perform outside his window.
He eventually lost his temper with this mirthful musical assault, exited his
front door, and went in search of a policeman:

> In the meantime the crowd of young children, urged on by their parents,
> and backed at a judicious distance by a set of vagabonds, forms quite a
> noisy mob, following me as I pass along, and shouting out rather uncom-
> plimentary epithets. When I turn round and survey my illustrious tail, it
> stops; if I move towards it, it recedes: the elder branches are then quiet—

sometimes they even retire, wishing perhaps to avoid my future recognition. The instant I turn, the shouting and the abuse are resumed, and the mob again follow at a respectful distance.[45]

Babbage vividly recounted his experience of turning to face the music: the threatening crowd encroached from behind, but grew quiet as soon he turned around—the mob disaggregated before his eyes, individual faces became unreadable. The problematic nature of identifying the mob resided in a split between an elusive visual reality and a seemingly irresistible aural channel. Sensation and ideology were conflated: the semantic freight of noise transferred from the social problem of street music onto the shouts and abuses of a mob. Babbage's implicit political aim was to bring this mob to heel. And this was what his opposition to street music implied: bolstering hegemony through a contest over sensory domains—what Jacques Rancière calls the always-conflicted "distribution of the sensible."[46]

I want to suggest that these sensory fluxions—patterns of perception that describe the contested field of politics—can be mapped locally, the walls of Babbage's Dorset Street home serving as a partition. Indoors, the elderly scientist labored on sketches for his engines, and the physical effort involved in this task defined his tangible hostility to street music:

> I claim no merit for this resistance; although I am quite aware that I am fighting the battle of every one of my countrymen who gains his subsistence by his intellectual labor. The simple reason for the course I have taken is, that however disagreeable it has been, it would have been still more painful to have given up a great and cherished object, already fully within my reach.[47]

When Babbage wrote these words, he had been designing calculating engines for over forty years. It is remarkable that he still entertained the prospect of completing his machine, long after the workshops around his home had shut down. Yet by 1864, his drive for "intellectual labor" on his engines had been substantially diverted into his campaign against street music. By expelling the intruders from around his home, he sought to calm the acoustic environment, thereby reestablishing conditions for steady concentration—which had now become an end in itself.

In attending to, and fighting against, the mechanical strains outside his windows, Babbage listened to the industrial city as though it were itself a machine, an autonomous social system that had taken on unstoppable, destructive force. The hoi polloi was polluting the atmosphere around his home, which ought to have been the city's scientific nervous center (or so he thought). Within these musically contaminated environs, the notion that "one-fourth part" of his working power had been "destroyed" takes on renewed significance. Reading on, we learn that this residential soundscape was entwined with a projected industrial-intellectual economy: "Twenty-five per cent is rather too large an additional income-tax upon the brain of the intellectual workers of this country, to be levied by permission of the Government, and squandered upon its most worthless classes."[48] Here again, Babbage's factory-like mind revealed its aggressively economizing contours: its obsessions with the productive limits of thought. And it is at these limits that noise emerges as an object of political concern, making perceptible the link between brainwork and street music:

> When the work to be done is proportioned to the powers of the mind engaged upon it, the painful effect of interruption is felt as deeply by the least intellectual as by the most highly gifted. The condition which determines the maximum of interruption is,—that the mind disturbed, however moderate its powers, shall be working up to its full stretch.[49]

Intruding via the auditory pathway, street music pushed Babbage's working mind into overdrive. Its sonic pressure *was* the social friction that resisted the progress of his machine—and, by extension, the development of the economy, the city and the nation as a whole.

AN ECONOMY OF LISTENING

In 1864, the year Babbage's "Street Nuisances" was published, Michael Bass's Bill for the Better Regulation of Street Music in the Metropolis was approved by Parliament. The latter meant that a street musician could be removed from the vicinity of any given home if there happened to be indoors someone "engaged in some serious occupation which required to be carried on without interruption."[50] Before Bass's amendment, only the presence of an invalid in the home permitted the property owner to have musicians removed by police officers; now reasonable cause also protected professionals working from home. With Babbage's experience before us, we might choose to pay special attention to Bass's carefully placed modifiers: his bill referred to "serious" occupations, which should be allowed to continue "without interruption"—qualifications that hint at pivotal nodes in the broader parliamentary debate. First among these were the competing claims of intellectual laborers versus workers in the street music trade; second, the basic (but nonetheless contested) assumption that silence was required for intellectual labor to be pursued. These contested points defined the terms of discussion and the polarized background that Babbage's extreme views had been instrumental in instigating. London's economy was apparently imperiled, the metropolitan environment being identified as a zone of contest, where political action was figured in terms of the struggle over the city's sensible domains.

We might, in conclusion, try to sketch anew these parliamentary debates over street music with our ears trained on these economic and sensory aspects. Babbage's eccentric positions on these themes can illuminate, precisely because of their offbeat emphases, the larger discussion over street nuisances. Yet this indirect synergy between individual action and official politics can best be illustrated not by the successful implementation of Bass's act in 1864, but instead through Lord Westmeath's failed attempt to introduce his Barrel Organ Suppression Bill five years earlier. The parliamentary transcripts pertaining to this document evoke an insurgent, yet for now ill-defined anti-street-music sentiment. Westmeath's opening deposition played on (the by now familiar) xenophobic fears, the argument being framed in terms of exploitative/exploited migrant worker-musicians:

> The persons who annoyed the inhabitants of London were, as their Lordships were aware, chiefly foreigners, and were brought over here by persons who made a profit of their earnings, allowing them only a bare subsistence. He admitted that hospitality was due to foreigners, but he denied that the peace and tranquility of the metropolis were to be sacrificed to their convenience and profit. . . . Several persons had objected to his Bill on the ground that it would deny the public the gratification of listening to the German bands. That was a mistake. A man could not keep on blowing a wind instrument for ever; but a barrel organ never tired; it was a nuisance which never ceased, and was an object of universal detestation; and it was the object of his Bill to suppress it.[51]

Westmeath's parataxis revealed a jumble of complaints: first, barrel organs were more odious than wind instruments; second, they were mechanical and unremitting; third, everyone hated them. Subsequent speakers would challenge each of these points. Yet taken together, Westmeath's objections added up to a particular ideology: in comparing German bands with (Italian) organ grinders, he sought to identify the industrial-scale nuisance with its mechanical means of production. In other words, street music was denounced for its machine power—a force elsewhere celebrated as the cornerstone of the British economy. Westmeath claimed that the sound of barrel organs "never ceased": much like capital flows theorized by Marx at the time, these instruments were allied to forces of circulation and production that had taken on a life of their own.[52] It was this systematic dimension of street music that also concerned Babbage; although, as I have argued, his complaint was more intricate than Westmeath's: he understood street music as a negative force that was in direct and sensuous conflict with the intellectual economy.

Westmeath's parliamentary opponents seized upon the idea of the economic vitality of street music.[53] Lord Lyndhurst, for example, charged that his learned friend had been blinkered to the wider economic and moral benefits of street music. These benefits were proved, Lyndhurst claimed, by the fact that the players were so often well-paid: "and if they were it could only be because their performances were agreeable to the humble proprietors of homes in that district, and [Lyndhurst] had as little doubt that it exercised a softening influence on their manners."[54] Invoking music's allegedly civilizing effects, he made the case for the propagation of mechanical music as a positive force. What is more, Lyndhurst saw no reason why such music should disturb those engaged in intellectual work—directly challenging Westmeath's (and Babbage's) implicit position that noise made concentration impossible. When pursuing his own mathematical studies in his London chambers, Lyndhurst explained, he trained himself to ignore a neighbor playing the violin, and had thus come to enjoy the more general benefits of greater resilience against widespread musical interruptions in the city. Such interruptions were, after all, an inevitable part of everyday life in a healthy industrial economy: Lyndhurst took pride (and masochistic pleasure?) in his ability to withstand disturbances; he encouraged his fellow lords and countrymen to demonstrate similar capacities for industry-resistant mental vigor.[55]

In the parliamentary debate, Lyndhurst was succeeded by Earl Granville, who seconded "every word that has fallen from the noble and learned Lord." However, Granville wanted to add another, more class-sensitive argument about the virtues of healthy, cheap street music:

> Only a very small proportion of the community were gifted with such exquisite ears that they could endure none but the most refined and costly music; and he could not see why, for the sake, perhaps, of some rich and highly-sensitive connoisseur, a whole neighborhood of poor people should be debarred of the innocent pleasure of listening to a barrel organ. The allusion to foreigners in the preamble of the Bill seemed to pander to an unworthy prejudice; and for his own part he infinitely preferred the

performances of a German band to the favored musicians alone exempted from the operation of the measure—namely, the sham base [*sic*] and falsetto singers, who trusted to the strength of their own lungs for their success, instead of having recourse to a much milder and more harmless instrument.[56]

As we have seen several times already—and is restated clearly here—the street music trade either caused offense or gave pleasure according to the business model being invoked—and the wider influence that that business could be presumed to have on the urban economy. For while Granville drew conclusions opposed to Westmeath's, they nonetheless concurred in their methods of assessing street music according to the sonic means of production: the "refined and costly music" represented by operatic voices were evaluated by the "strength of [the singer's] lungs," while the "innocent pleasure" of the patronized poor was increased by the "much milder and more harmless" entertainment represented by the barrel organ. In other words, street music was neither good nor bad in itself, but it stood for sheer sonic power. The act of hearing street music was split by a prevailing ideology, which encouraged politically minded Londoners both to listen *to* street music and also *through* it, for the human and/or machine labor involved in producing sonic energy. In mid-nineteenth-century London listening to street music thus came to mean, at least in part, listening to the circulation of sound within the metropolitan economy.

This view, encapsulated by the 1859 parliamentary debate, stands in naked contrast to Babbage's, which understood street music as a forcible drag on the economy. But the oppositional context represented by Lyndhurst and Granville (and other lords besides) helps define Babbage's campaign against street music as a political act: as the struggle of a particular, peculiar activist in the name of a utilitarian, tightly disciplined metropolitan order. Babbage promulgated an alternative, more ruthlessly industrial ideal for the economy in which the division of labor—that of workers, both manual and intellectual—might continue unimpeded by street music, thus reaching toward its productive maximum. What sustained this political configuration of musicians and machines, were the curious objects brought into fleeting aural contact at Dorset Street: Babbage's prospective calculating engines inside and the mechanized organ grinders without.[57] As engine and barrel organ were brought into proximity, Babbage morphed into a distinctive political actor: a prototype of an industrial human being tethered to the economy by the ear.

We have encountered Babbage's android listener once before in this essay: it is the deskilled worker imagined through his calculating machines, which would have required of their operators *merely* to listen to numbers communicated by bells. These industrial, industrious sounds emerging from Babbage's home form a surprising counterpoint to the noisy streets beyond, calling attention to the untold ways in which Victorian capitalism interacted with listeners, and with modes of listening, in (and to) the city.[58] And it is, ultimately, against these broader vistas of urban, industrialized perception that Babbage's campaign against street music most powerfully signifies: against a polarized urban musical culture in which street music was defined by its distant relation to what was going on inside, in the rarified atmosphere of opera or concert or chamber music.[59] There is an unmistakable symmetry between concert audiences falling silent during the nineteenth century and Babbage's anguished efforts to preserve the equilibrium of his acoustic environment. Music indoors provided the unconscious blueprint for music outdoors, suggesting to Babbage, as to the intellectual laborers who stood behind Bass's bill, the utopia of an ambient texture in which patterns of attention were everywhere put to intelligent, productive use.

NOTES

1. Lord (John Fletcher) Moulton's speech, "The Invention of Logarithms, Its Development and Growth," was published in the conference's proceedings: Cargill Gilston Knott, ed., *Napier: Tercentenary Memorial Volume* (London: Longmans, Green & Co., 1915), 1–32. In his foreword, Knott noted that the conference had been "attended [by friendly men and women] whose nationalities were fated to be in the grip of war before a week had passed." During the First World War, Moulton (1844–1921)—a barrister, mathematician, politician, and renowned polymath—would become director general of the Explosives Department.

2. Emphasizing the role of the popular music industry in shaping legal and institutional definitions of intellectual property, Adrian Johns traces the growth of antipiracy measures in Britain to the early twentieth century; see Johns, *Piracy: The Intellectual Property Wars from Gutenberg to Gates* (Chicago: University of Chicago Press, 2009), 327–56.

3. Moulton, "The Invention of Logarithms," 24.

4. Completing the narrative of scientific catastrophe, Moulton went on: "When he died a few years later, not only had he constructed no machine, but the verdict of a jury of kind and sympathetic scientific men who were deputed to pronounce upon what he had left behind him, either in papers or mechanism, was that everything was too incomplete to be capable of being put to any useful purpose" ibid., 20). Martin Campbell-Kelly cites this passage in his introduction to *Charles Babbage: Passages from the Life of a Philosopher* (London: Pickering, 1994), 34; Campbell-Kelly cautions against a too-literal reading of Moulton here, underscoring the likelihood of embellishment in his recollection of the "facts."

5. The public funding of science did not become common until the twentieth century; however, scientific institutions, such as the British Society for the Advancement of

Science—founded in 1831 by, among others, Babbage himself—lobbied the government for financial sponsorship. See Peter J. Bowler and Iwan Rhys Morus, *Making Science Modern: A Historical Survey* (Chicago: University of Chicago Press, 2010), 329–37.

6. In 1991, to celebrate the centenary of Babbage's birth, a working difference engine no. 2—based almost entirely on his original designs—was built and put on display at London's Science Museum.

7. A newspaper article published in 1834 first drew attention to interruptions to Babbage's work on the machine; see D. Lardner, "Babbage's Calculating Engine," *Edinburgh Review* 59 (1834): 263–327; reprinted in Charles Babbage, *The Works of Charles Babbage*, 11 vols., ed. Martin Campbell-Kelly (New York: New York University Press, 1989), 2:118–86. The press remained broadly sympathetic to Babbage's aims, trying hard to explain the device to the public during the 1830s, '40s, and '50s. However, his work on the perpetually nonexistent "calculating machine" also became an object of ridicule. See Campbell-Kelly, introduction to Babbage, *Charles Babbage: Passages from the Life of a Philosopher* (New Brunswick: Rutgers University Press, 1994), 1–34.

8. Editorial, *Morning Chronicle* (London), August 25, 1856. Five years earlier, Babbage had fallen into a dispute with the same newspaper: it was his policy to fight back against satire. In 1851, he leveled an accusation of calumny against the newspaper, over claims he had completely abandoned the difference engine (in favor of the analytical engine). The *Morning Chronicle* hit back that Babbage's claim was "a most extraordinary instance of the hallucinations into which men of genius may be hurried by irritability"; see "Mr. Babbage," *Morning Chronicle* (London), July 4, 1851.

9. Charles Babbage's "Street Nuisances" was published a few weeks ahead of (and subsequently included in) his autobiography, *Passages from the Life of a Philosopher* (London: Longman, Green, Longman, Roberts and Green, 1864). Babbage seems to have set a precedent for anti–street-nuisance tracts; see Henry Renshaw, *The Nuisance of Street Music or, A Plea for the Sick, the Sensitive, And the Studious. By A London Physician.* (London: 365 Strand, 1869), and W. C. Day, C.B., *Street Nuisances; A Letter to Colonel E. Y. W. Henderson, Her Majesty's Chief Commissioner of Police, on the Condition of the Strand and Other Leading Thoroughfares of the Metropolis* (London: William Tweedie, 1871).

10. John M. Picker, *Victorian Soundscapes* (Oxford: Oxford University Press, 2003), 41–45.

11. Babbage, *Passages from the Life of a Philosopher*, 345.

12. In this article, Babbage mentioned an engine "which is just finished" that could calculate to two orders of difference. See Charles Babbage, "A Note Respecting the Application of Machinery to the Calculation of Astronomical Tables" *Memoirs of the Astronomical Society* 1 (1822): 309, reprinted in Babbage, *The Works of Charles Babbage*, 2:3–4.

13. "A Letter to Sir Humphry Davy, Bart, President of the Royal Society, On the Application of Machinery to the Purpose of Calculating and Printing Mathematical Tables" (1822), reprinted in Babbage, *The Works of Charles Babbage*, 2:6.

14. In his initial letter to Davy, Babbage stressed that the machine would be a long-term investment: "It must be obtained at a very considerable expense, which would not probably be replaced, by the works it might produce, for a long period of time, and which is an undertaking I should feel unwilling to commence, as altogether foreign to my habits and pursuits" (ibid., 2:14).

15. Lorraine Daston, "Enlightenment Calculations," *Critical Inquiry* 21, no. 1 (1994): 186.

16. See F. Baily, "On Mr Babbage's New Machine for Calculating and Printing Mathematical and Astronomical Tables," *Astronomische Nachrichten* 46 (1823): col. 409–22, reprinted in Babbage, *The Works of Charles Babbage*, 2:44–56.

17. Daston, "Enlightenment Calculations," 200; see also Bowler and Morus, *Making Science Modern*, 397.

18. William J. Ashworth, "Memory, Efficiency and Symbolic Analysis: Charles Babbage, John Herschel, and the Industrial Mind," *Isis* 87 (1996): 629–53.

19. See Simon Schaffer, "Babbage's Dancer and the Impresarios of Mechanism," in *Cultural Babbage: Technology, Time and Invention* (London: Faber & Faber, 1996), 53–80.

20. See Anthony Hyman, *Charles Babbage: Pioneer of the Computer* (Princeton, NJ: Princeton University Press, 1985), 128.

21. Babbage, *Passages from the Life of a Philosopher*, 82. During another hiatus from work on the engine in 1841, Babbage wrote the script for a ballet involving complex lighting effects; he even entered into discussions with Benjamin Lumley, manager of the Italian Opera House; see Ivor Guest, "Babbage's Ballet," *Ballet* 5, no. 4 (1948): 51–56.

22. "The first and great cause of [the difference engine's] discontinuance was the inordinately extravagant demands of the person whom I had employed to construct it for the Government. Even this might, perhaps, by great exertions and sacrifices, have been surmounted" (Babbage, *Passages from the Life of a Philosopher*, 449).

23. Schaffer, "Babbage's Dancer"; see also Simon Schaffer, "Babbage's Intelligence: Calculating Engines and the Factory System," *Critical Inquiry* 21 (1994): 203–27.

24. Schaffer, "Babbage's Dancer," 70.

25. On automata and pseudo-automata in early nineteenth-century London, see also the essays by Myles Jackson and Melissa Dickson in this volume.

26. Charles Babbage, *On the Economy of Machines and Manufactures* (London: Charles Knight, 1832), 160–61.

27. There is another set of associations between Babbage and "difference" stemming from his aggressive promotion of Leibniz's d-notation in differential calculus, in which "dx" represents the infinitesimally small increment (that is, difference) of some variable x. While now commonplace, d-notation was hotly contested among mathematical circles in the 1810s: it was associated with the radicalism and lawlessness of the French Revolution. The cause of d-notation was taken up by the young Babbage under the aegis of his Analytical Society, run by students at Cambridge University. See Andrew Warwick, *Masters of Theory: Cambridge and the Rise of Mathematical Physics* (Chicago: University of Chicago Press, 2003), 67–68.

28. Babbage claimed that, before exhibiting the machine, he had never attached bells to it: he had not previously considered "the power it thus possessed to be of any practical utility" (Babbage, *Passages from the Life of a Philosopher*, 66). This is despite the previously mentioned discussion of bells in *On the Economy of Machines and Manufactures*, and further previous references to the projected use of bells in the analytical engine (see, e.g., Babbage, *Passages from the Life of a Philosopher*, 119). Shortly after Babbage's death, his son recalled another purpose for bells being attached to the difference engine: "The engine would also be set to stop itself as soon as it had completed such a number of calculations as would be true to the last figure printed, this number having been ascertained by the operator beforehand: it would then ring a bell to draw attention to its need of a fresh difference, and throw itself out of gear so as to stop the work and prevent the possibility of any inattention on the part of the operator allowing an error to creep in"; see Benjamin

Herschel Babbage, "Calculation of Tables Having No Constant Difference" (1872); reprinted in Babbage, *The Works of Charles Babbage*, 2:232.

29. Babbage, *Passages from the Life of a Philosopher*, 65–66.

30. Schaffer has drawn attention to the intimate connection between the illusion of artificial intelligence in Babbage's machines and cultures of scientific display in early nineteenth-century London; see Schaffer, "Babbage's Dancer." For a recent treatment of the topic of science and spectacle, see Joe Kember, John Plunkett, and Jill A. Sullivan, eds., *Popular Exhibitions, Science and Showmanship, 1840–1910* (London: Pickering and Chatto, 2010).

31. Picker, *Victorian Soundscapes*, 41–81. Picker discusses the way in which street music served as a foil for the construction of middle-class professional, male identity in mid-Victorian London for a range of writers working from home. By making the streets quiet and thus "domesticating" them, Picker argues, these writers sought to establish their dominance over—and to articulate their sonically marked difference from—both the lower orders of society and foreign street musicians. My argument in this essay is slightly different; by focusing on Babbage—whose ideas were eccentric, but for that reason can illuminate broader common ground—I want to underscore the economic basis of contemporary debate. In other words, despite the huge difference in status between street musicians and professionals (such as scientists), such class distinctions threatened to be undermined by an emerging social hermeneutics, whereby both could be understood to be productive workers within the same economy—different kinds of workers (but workers nonetheless) who were thrown into sensible contact and competition through their copresence in space.

32. Policemen were reluctant to arrest street musicians, perhaps in part because the Met enjoyed a lively musical component: see Rachel Cowgill, "On the Beat: the Victorian Policeman as Musician," in *Victorian Soundscapes Revisited*, ed. Martin Hewitt and Rachel Cowgill (Leeds, UK: University Print Services, 2007), 191–214. Babbage's correspondence with Westmeath and Bass can be found in the British Library, MS 37197, ff. 434, 455, 459 and MS 37198, ff. 513, 529.

33. Babbage, *Passages from the Life of a Philosopher*, 352–53.

34. See John E. Zucchi, *Little Slaves of the Harp: Italian Child Street Musicians in Nineteenth-Century Paris, London and New York* (London: McGill-Queen's University Press, 1992), 1–41, 76–110. See also: Henry Mayhew's account, "Street Musicians," in *London Labour and the London Poor: The Condition and Earnings of Those That Will Work, Cannot Work, and Will Not Work* (London: Charles Griffin & Co., 1851), 3:168–99.

35. See, e.g., Mayhew's account, "Organ Man, With Flute Harmonicon"; *London Labour and the London Poor*, 3:184.

36. Arthur W. J. G. Ord-Hume, "Barrel Piano," *Grove Music Online: Oxford Music Online* (Oxford University Press), accessed March 8, 2014, http://www.oxfordmusiconline .com/subscriber/article/grove/music/02112.

37. Babbage, *Passages from the Life of a Philosopher*, 338.

38. On London's popular music economy, see Derek B. Scott, *The Sounds of the Metropolis: The 19th-Century Popular Music Revolution in London, New York, Paris, and Vienna* (Oxford: Oxford University Press, 2008), 15–37.

39. Babbage, *Passages from the Life of a Philosopher*, 338.

40. The tendency for Londoners to view their streets (and, by association, street music) as immoral was widespread: see Heather Shore, "Mean Streets: Criminality, Immorality and the Street in Early Nineteenth-Century London," in *The Streets of London: From the*

Great Fire to the Great Stink, eds. Tim Hitchcock and Heather Shore (London: Rivers Oram, 2003), 151–64.

41. Along similar lines, Brenda Assael explores the implications of noise as sound out of place; see her "Music in the Air: Noise, Performers and the Contest over the Streets in Mid-Victorian Britain," in Hitchcock and Shore, *The Streets of London*, 183–207.

42. The full table was published in Michael Bass's *Street Music in the Metropolis* (London: John Murray, Albemare Street, 1864), 20–22.

43. Babbage, *Passages from the Life of a Philosopher*, 347.

44. Ibid., 345.

45. Ibid., 349.

46. Jacques Rancière, *The Politics of Aesthetics: The Distribution of the Sensible* (London: Continuum, 2006), 7–14.

47. Babbage, *Passages from the Life of a Philosopher*, 345.

48. Ibid.

49. Ibid., 346.

50. "Leave," *Hansard* (House of Commons debate held on May 3, 1864), vol. 174, 2116–19. In the convoluted wording of the bill itself, "any Householder within the Metropolitan Police District, personally, or by his Servant, or by any Police Constable, may require any Street Musician or Singer to depart from the Neighbourhood of the House of such Householder, on account of the Illness, or on account of the Interruption of the ordinary occupations or Pursuits of any Inmate of such House, or for other reasonable or sufficient Cause; and every Person who shall sound or play upon any Musical Instrument or shall sing in any Thoroughfare or public Place near any such House after being so required to depart, shall be liable to a Penalty not more than *Forty Shillings*, or, in the Discretion of the Magistrate before whom he shall be convicted, may be imprisoned for any Time not more than Three Days." "Act for the Better Regulation of Street Music in the Metropolis," ordered to be printed by the House of Commons on May 4, 1864.

51. "Second Reading: Negatived," *Hansard* (House of Lords debate held on April 29, 1858), vol. 149, 1925–30. See also: *Hansard* (House of Lords debate held on April 20, 1858), vol. 149, 1351–53.

52. On Marx as a reader of Babbage, see Nathan Rosenberg, *Exploring the Black Box: Technology, Economics, and History* (Cambridge: Cambridge University Press), 24–46.

53. James Winter speculates that many voices in parliament stood by street music because it represented a political safety valve: "street music provided a cultural meeting ground for almost every segment of society . . . . [It] was one of the few aspects of urban life that just about everyone could enjoy, including, probably, many of those philosophers, artists, composers, scientists, and men of letters who signed Bass's petition." See James Winter, *London's Teeming Streets, 1830–1914* (London: Routledge, 1993), 77.

54. "Second Reading: Negatived," 1928.

55. On the Victorian manliness and mathematical studies, see Warwick, *Masters of Theory*, 176–226.

56. "Second Reading: Negatived," 1929.

57. See Bruno Latour, *Reassembling the Social: An Introduction to Actor-Network-Theory* (Oxford: Oxford University Press, 2005), 63–70.

58. Jonathan Crary discusses the synergy between industrialized labor and the attentive aesthetics of nineteenth-century visual culture in *Suspensions of Perception: Attention, Spectacle, and Modern Culture* (Cambridge, MA: MIT Press, 2001).

59. References to concert music are rare in Babbage's writings, but in his autobiography he hints at a connection between mechanization and attentive listening in an account of a concert held at Hanover Square Rooms in the early 1840s—one of the orchestral and choral extravaganzas organized by the elite Society for Ancient Music: "Soon after I had taken my seat at the concert, I perceived Lady Essex at a short distance from me. Knowing well her exquisitely sensitive taste, I readily perceived by the expression of her countenance, as well as by the slight and almost involuntary movement of the hand, or even of a finger, those passages which gave her most delight. These quiet indications, unobserved by my friends, formed the electric wire by which I directed the expressions of my own countenance and the very modest applause I thought it prudent to develop" (Babbage, *Passages from the Life of a Philosopher*, 427–28).

CHAPTER NINE

Hearing Things: Musical Objects at the 1851 Great Exhibition

FLORA WILLSON

Without things, we would stop talking.
—Lorraine Daston, *Things That Talk*

In late May 1851, Martin Cawood—wealthy Yorkshireman, brass and iron founder, and amateur musician—visited the Great Exhibition in London. What he experienced there, as he described it in a letter to the editors of the *Leeds Mercury*, was "an incessant whirl of hustle and bustle."[1] As did so many other commentators and eyewitnesses, he praised the regular attendance of Queen Victoria; gushed that the central nave was "heavenlike for its elegance and transparency"; and admired the "fairylike splendor" of so many richly attired visitors converging in such a magnificent space. Yet he also offered a more unusual response, outlining the sensory effects of the Exhibition on the roaming visitor:

> Some elegant work of art immediately attracts his eye—but before that organ of vision can dwell, even for a moment, upon its beauties, the ear is arrested by sounds foreign to the accustomed tones, and French, German, Dutch, Spanish, and other languages assail it in a strange medley of sounds. Attracted for a moment by this, he turns round to look at his next neighbor, and in doing so his eye is again caught by some new object. Forgetful of the first, he at once rushes to the second. Yet he grasps as it were at a shadow. . . . His ear is assailed with the pealing strains of an organ, or the brilliant tones of a piano, and he rushes to the place from whence they proceed. Here again he is disappointed. Some scores of anxious listeners surround the performer, and the buzz and noise around lead him to endeavor to reach some less frequented place. In vain he attempts

it, and seeks the nearest seat, only to search for a vacant place in vain.
Bewildered, perplexed, and confused, his brain becomes overpowered.[2]

In Cawood's account, the Exhibition visitor is overwhelmed not so much by
the beauty of his surroundings as by the sheer variety of sights and sounds
competing for his attention. His eyes and ears are constantly drawn in different directions: visual objects attract him and fade from sight in a constant
stream; foreign tongues mingle confusingly; music heard from afar turns to
noise once tracked to its source. Sustained concentration in such an environment was little short of impossible. Small wonder that Cawood describes the
listeners crowded around a musical performer as "anxious."

Such descriptions of perceptual overload and auditory anxiety provide a
fitting beginning to a chapter about music at the Great Exhibition. Indeed, as
I attempt to trace its place in and contribution to that famous event, music's
absence—its tendency to dissipate, to dematerialize—will be as significant
as its more stable or formal manifestations. If the matter of concern here is
"music," I take my conception of what "music" might encompass from the
evidence of the Exhibition itself, and so will consider the role and extent of
musical performance alongside a more unusual perspective: one that begins
by studying the classification of musical objects on display. This perspective, which does not take issues of musical ontology for granted, in turn
raises much larger questions, even extending to what constituted a musical
object in London, ca. 1851. My broader aim is to interact with a theoretical
discourse long established in musicology: the debate still accumulating—
almost thirty years after its first appearance—around Lydia Goehr's book
The Imaginary Museum of Musical Works. The notion of what Goehr called
the "work-concept" is now commonplace, as is her "central claim" about how
that notion became regulative in the years around 1800.[3] Yet, as this chapter will seek to demonstrate, the Great Exhibition's classification of musical
objects (and particularly its classification of musical instruments) might be
productively understood as a putative, literal materialization of an epistemology of music indebted to and coterminous with the emergence of the
work-concept, as set out by Goehr. To put this another way, while much of
the (vast) literature on the work-concept emerging since Goehr's book has
responded to her ideas about the "musical work," one could—and perhaps
should—press harder on the "museum" in her now-iconic title.

THE "GREAT EXHIBITION OF THINGS"

The Great Exhibition of the Works of Industry of All Nations, as it was officially known, opened in London's Hyde Park on May 1, 1851. It was housed in
a vast, purpose-built structure of glass and iron that had already been christened the "Crystal Palace" by *Punch* magazine. This technologically innova-
tive venue was the most visible expression of the Exhibition's basic ideology
and purpose: its objective (according to one contemporary announcement)
was no less than to "chart the progress of mankind."[4] As such, the event
was to epitomize the "Age of Machinery" so famously identified—and regret-
ted—by Thomas Carlyle in 1829.[5] In 1851, however, Carlyle's machine age was
not merely represented but also explicitly celebrated; its "whole undivided
might" was now on show, gesturing toward an idealized industrial future.

Notwithstanding the Exhibition's claims to measure the distance traveled from its more technologically primitive past and the "watershed" rhetoric that consequently accumulated around it in 1851, such an event was
not without precedent. National and local industrial exhibitions had been
mounted with considerable success in France and Britain during previous decades; the Birmingham Exhibition of Manufactures and Art, which opened
on September 3, 1849, was one particularly important precursor.[6] The 1851
Exhibition did, however, depart from previous exhibitions in three significant ways. First, it constituted the earliest such event mounted in Britain
under official government auspices. It was organized by a Royal Commission enthusiastically led by Prince Albert, who hoped thereby to foster both
"competition and encouragement."[7] (French exhibitions had been explicitly
national institutions since the first event in 1798.) Second, it was the first
attempt to stage an exhibition that was international in scope. Given the
biases of the era in general, and of the British imperial project in particular,
it is hardly surprising that the promised participation of "All Nations" was
downgraded in the *Official Catalogue* to "almost the whole of the civilized
nations of the globe."[8] The Royal Commission could nonetheless boast that
it had accommodated approximately fifteen thousand exhibitors: roughly
half—and thus half of the floor-space of the building—were either British
or from British colonies; the remainder hailed, according to the *Catalogue*,
from "over forty foreign countries."[9] This international purview in turn enabled the Exhibition's third, and most important innovation: as *Eliza Cook's
Journal* explained, the event had the effect of "laying out the industrial progress of the world, as it were, on a race-course, and indicating the positions
which the various countries occupy in respect to each other."[10] Thus not only
was the Exhibition a celebration of its "Age of Machinery" and a materializa-

tion of "industrial progress" for all to see; it was also, crucially, a means by which to compare the relative progress made by different countries. As such, the Exhibition was to be "to industry what galleries of painting and sculpture are to art—what a library is to literature—what a museum is to science—what a zoological and botanical garden is to natural history."[11]

There are rich seams of irony running through such rhetoric, widespread as it was in the run-up to the Exhibition's opening. What interests me above all is how the Exhibition was understood—at least in theory—to encapsulate in a static, viewable form an ongoing process of change. It is symptomatic that the first so-called histories of the Exhibition were published long before its doors closed in October 1851. The event was monumentalized in something like real time, its own fundamentally historiographical rationale—its synthetic staging of "progress"—rendered newsworthy in and of itself. It was precisely for the purposes of monumentalization that such a prestigious literary magazine as the *Athenaeum* could, with unblushing confidence, declare the Exhibition "the great historical centre of the nineteenth century. In an age which has been full of wild revolutions, great deeds, stirring events, it is the greatest deed and event of all."[12] This view plays fast and loose with history. Writing at the close of an exhibition intended to demonstrate and celebrate industrial progress, the *Athenaeum* relocated the cause for celebration to the Crystal Palace itself. The Great Exhibition was no longer acclaimed simply as a collection of exhibits in space, but as a Great Event—even *The Great Event*—in a century of which forty-nine years remained.

Such historiographical panache is striking from the pen of a mid-nineteenth-century writer. What is more thought provoking, however, is that recent writers in a variety of disciplines have continued where such contemporary commentary left off. The Exhibition has been repeatedly positioned as a defining moment in nineteenth-century science, industry, and culture; above all, as a symbol of the final, unstoppable encroachment of industrial modernity into culture and everyday life, a juncture literally materialized for all to see. This fever-pitched claim is maintained even by the most recent (and declaredly revisionist) essay collection to appear on the topic in literary studies, its editors suggesting that although "officially promoted as a comprehensive representation of global progress, the Exhibition also became an unofficial forum on the meanings of modernity."[13]

These are high stakes. Yet amid such observations of the Exhibition's central position as viewed from a variety of disciplinary perspectives, the contribution of musicology has been strangely muted. One might identify various reasons for such neglect. Most significant among them is the fact that musical performance has long been thought peripheral to, if not entirely lacking from, the events of the 1851 Exhibition.[14] In this historical reading, music was put to the most perfunctory of uses in the opening ceremony before being silenced for the duration of the Exhibition itself. Only when the relocated, remodeled Crystal Palace opened in Sydenham in 1854 did the building become a venue for musical performance—most famously for the "monster" Handel festivals launched in 1857 and mounted triennially between 1859 and 1926. Michael Musgrave's study of the Crystal Palace has done more than any other to examine all aspects of music making in and around the iconic building, but even Musgrave passes quickly over the months of the Great Exhibition itself. There, he states, "music's role was restricted to the mechanical," and "musical performance on a broader scale had no place as such"; his account focuses instead on the building's long and reverberant afterlife.[15]

All this argument might seem to be the prelude for an elaborate act of revision: an unearthing of forgotten performance at the 1851 Exhibition and an attendant claim for music restored to its rightful position. Or perhaps, following the lead of scholars elsewhere in the humanities, one might call for musicology's immediate and whole-hearted embrace of the Great Exhibition as a glittering index of modernity. But such approaches would be at least as problematic in this field as they have proved elsewhere. What might be more valuable is to query the widespread assumption that musical performance was absented from much of the Exhibition; there may be room in our histories for other manifestations of musical experience that we might detect there. The Exhibition's classification of musical objects provides a productive starting point for this inquiry. Its displays—of organs, pianos, and other finished instruments, as well as of numerous internal mechanisms and components essential to the workings of such instruments, and (just as significant) of printed scores—constituted the most obvious way in which "music" was incorporated into the world presented at the Crystal Palace.

The decision to include musical items within an event that proclaimed itself as a celebration of "industrial works" seems not to have attracted the same controversy as did the Exhibition's single so-called Fine Arts Court. (The latter paradoxically excluded most fine arts products in order to allow room for those demonstrating new methods and media; and its mediocre contributions were derided by many as an embarrassment to British art.) Musical instruments nevertheless proved difficult to categorize in an exhibition dedicated explicitly to industrial progress and innovation. Worse still, they threatened to disrupt a carefully maintained distinction between art and industry. To examine how the musical displays were sorted and then staged, as well as how they were understood by their viewing publics, thus seems one clear way to excavate aspects of musical thought around 1851 — at least as such thought was manifested outside of London's elite musical institutions.

THE CLASSIFYING IMAGINATION

The ordering and distribution of musical objects at the Exhibition can be understood, in the first instance, as only one element (albeit an especially problematic one) within a much larger taxonomic project. The Exhibition was, by all accounts, the century's highest-profile staging of systematic classification to date. As such, it took place against the intellectual backdrop of a theoretical debate about classification that raged during the early nineteenth century in the field then known as natural history, and which requires a brief excursus here.

Much has been made of the emergence of natural history. Its changing organization of knowledge — and the shifting political, social, and scientific priorities on which such knowledge depended — has proven fertile ground for discussion of the history of the notion of "order" in Western culture. The resulting narratives are, inevitably, on the grandest scale; none more so than Michel Foucault's iconic *The Order of Things* (first published in 1966).[16] Foucault's virtuosic sketch of a gradual separation of "the animal itself" from the knowledge handed down about it (what he calls "animal semantics") is well known; more significant in the present context, though, is his subsequent identification of natural history's emergence from a gap opened up between words and things, as accumulated knowledge of a given object came to be understood as a mode of representation rather than an intrinsic part of that object. The epistemological revolution described by Foucault brought nothing less than new descriptive orders — new ways of knowing the world.[17] And although — as is characteristic of grand narratives — this particular revolution can boast neither a definite beginning nor end, it is clear that, during the eighteenth and early nineteenth centuries, naturalists gradually sought to distance themselves from what they considered a "disorderly past," instead promoting a newly systematic present of tables and diagrams.[18] The author of this brave new world — frequently proclaimed the "father of taxonomy" — was the Swedish naturalist Carl Linnaeus, whose *Systema Naturae* (first published in 1735, and subsequently much revised) is usually cited as the first instance of systematic classification.[19] Meanwhile Linnaean nomenclature, by which every organism is designated by a name created from two Latinate words (the first its genus, the second its species), largely persisted through the nineteenth century's taxonomic frenzies and is still, famously, in use today.

Yet the philosophy—what Harriet Ritvo has called "the classifying imagination"—underlying Linnaeus's system was backward-looking even when it first appeared; and its adherence to an Aristotelian conception of the innate "natures" of organisms led to criticism of its apparently "artificial classifications."[20] The early decades of the nineteenth century saw repeated calls for a rival mode of classification, one that might take into account a broader range of data about the object being classified and thus generate taxonomic categories based on a more general understanding of nature as a whole. This gave rise to what the great evolutionary biologist Ernst Mayr identified as the "empirical approach" to taxonomy, which abandoned *a priori* considerations in favor of a supposedly unbiased assessment of the totality of an organism's characteristics.[21]

Such self-declared empiricism was by no means the end of the story; according to most versions, by the mid-nineteenth century Charles Darwin was preparing to spark the next epistemological revolution with his *Origin of Species* in 1859. But, for the purposes of this chapter, what is most significant in the interminable nineteenth-century debates about classification is the constant tension between "natural" versus "artificial" categories. Although the value of a systematic approach itself remained undisputed, two basic taxonomic modes were at loggerheads at midcentury: one understood classification to be following the unified plan of a divine Creator, with all individual taxonomic categories as variants of a single underlying type; the other maintained that all such categories were necessarily artificial—as Mayr put it, "the arbitrary products of the ordering human mind"—despite the fact that many *taxa* were empirically found to be natural. Taxonomic categories were, in short, disconcertingly flexible, shifting all too easily, and ever prone to blurring.

We might return at this point to the predictions made ahead of the Great Exhibition's opening, that it would (and should) be what "a museum is to science—what a zoological and botanical garden is to natural history" while also functioning as "a race-course" on which nations might meet in peaceful competition. The vast spaces of the Crystal Palace had not only to display a "universe" of objects in an orderly fashion; they also had to celebrate and make visible the forward march of progress.[22] To borrow Tony Bennett's phrase from his classic essay "The Exhibitionary Complex," the Great Exhibition pivoted on "two new historical times—national and universal."[23] That is to say, the event enabled the comparison of individual nations' claims to

industrial modernity via a single, universal measure of progress, tailor-made
to the strengths of Great Britain herself. In other words, the flexibility of prevailing taxonomic systems could be employed to the classifier's own political advantage.

Both contemporary commentators and more recent accounts have reported at length on the classificatory system employed at the Exhibition. Its
rigor and sheer orderliness were celebrated as a particularly British trait—an
impression reinforced by the fact that, following drawn-out arguments about
modes of classification within the Royal Commission, Albert's original plan
for a universal system of classification according to object type had been
abandoned as impractical. Seeking as it did to display similar objects together
regardless of national origin (an optimistic embrace of the principles of free
trade and an attempt to represent the "evolutionary" stages of industrial production), Albert was eventually overruled as a result of concerns that the
staggered arrival of objects from across the globe would render classification
impossible until all were in situ. It was thus decided to implement a dual system: on the largest scale, the Exhibition would be organized by country, with
the British and colonial exhibits positioned in the Western part of the building and the rest of the world in the East. The British and colonial exhibitors
were then subject to further triage according to a taxonomy much debated
but eventually fixed by Leon Playfair, a professional administrator-turned-
Liberal member of Parliament, in consultation with the British manufacturers themselves (the rest of the world was left to organize itself as it saw fit).[24]
The resulting classificatory system was split into four broad categories, the
progression of which hinted at the "evolutionary" aspects of Albert's original model—Raw Materials, Machinery, Manufactures, and Fine Arts—and
which were, in turn, divided into thirty smaller subcategories.

Understood thus, as an event of lasting significance for an emerging "classifying imagination," the Great Exhibition can be seen to harbor conflict at
its very heart. Despite frequent comparisons between the Crystal Palace and
various cathedrals, the Exhibition itself took as its conceptual foundation the
secular idea that (as literary scholar Thomas Richards put it) "all human life
and cultural endeavor could be fully represented by exhibiting manufactured
articles."[25] This same idea evidently generated the critical trope of a visit to
the Exhibition as a world tour in miniature. The world on view at the Crystal Palace was one unequivocally populated by commodities. What is more,
and although—as Richards has rightly observed—the Exhibition was torn
between the functions of the museum (in some sense another religious institution) and those of the marketplace, it was the latter that dictated its modes
of display (even in the much-contested absence of price tags). In the words of

the *New Monthly Magazine*, the Exhibition was intended to "convey . . . universal palpable truths in the most efficient way, in the smallest given time."[26] Whether museum or marketplace, that is, the Exhibition operated within a strictly capitalist economy of information, one in which efficiency was key. To put this another way: the Great Exhibition was understood (and disputed) from its early stages as an unprecedented celebration of commodities. It was, as Richards has described it, a "Great Exhibition of Things."[27]

That musical "things" featured in this hymn to the material raises difficult questions—not least in view of the particular taxonomic difficulties they generated as a category. Although they were present in the Royal Commission's earliest lists of objects to be included in the Exhibition, musical instruments initially appeared in both the Manufactures and Machinery sections.[28] In many cases the differentiation between the placement of objects in Raw Materials and in Manufactures was the subject of considerable debate: this happened with steel, for instance, and with leather.[29] Yet musical instruments were distinguished by their crossing of a different boundary: between items of interest owing to their finished state (and thus, in this case, considered to have aesthetic worth as objects to be *looked* at) and mechanical devices valued because of what they could generate—because of their capacity, quite literally, to make music. What is more, we might even attribute this particular collapse of taxonomic categories to the fact that the musical instruments on display in the Crystal Palace gestured toward two distinct types of musical object. One typology involved the capacity of musical instruments (as machines) to produce sound and thus stood in some sense for the performance of musical works, grounded in the claims to immortality of elite culture; the other involved evaluating the value of a musical exhibit on the basis of its beauty as a material object, its aesthetic or pleasurable qualities deriving less from its musical affordances than from its decorative traits and assessments of its particular contribution to recent industrial innovation.

What is most striking about music's material presence in the Crystal Palace, however, is how little it appears to have been discussed by those organizing the Exhibition. The minutes of the Royal Commission's meetings in the almost two years leading up to its opening reveal very little about what those in charge of sorting these musical things thought music *was*. For all that it is clear that the Commission was closely concerned with the classification of exhibits in general, and with criteria for differentiating raw materials and manufactured products in particular, there is no trace of debate about the status of musical instruments. When the Crystal Palace opened in May 1851, music was represented explicitly in Class X, a subsection of Machinery that housed "Philosophical, Musical, Horological and Surgical Instruments." Thus the place of metronomes, flutes, opera glasses, and pianos in the Great Exhibition's microcosm was located only meters from the ear trumpets, sextants, clocks, and—most thought provoking of all—armies of artificial noses, legs, and teeth. This unholy jumble of objects did not go unremarked at the time. Indeed, it is difficult not to be intrigued by the juxtapositions produced (however unwittingly) by these dismantled "human" bodies—bodies displayed literally as machinery, literally objectified—and the complete "ensemble" (of an entire orchestra contained within a single instrument) embodied by the pianos, for instance.[30] Perhaps most striking of all was that one of the star musical attractions, Henry Willis's "monster organ"—an instrument also nicknamed "the Leviathan"—which boasted "the largest swell in Europe," was located in the Western gallery, and so was placed next to a model man, five feet tall but apparently capable of expanding to the height of six foot, eight inches. Again, the peculiar juxtaposition was noted, one guide seeing fit to "assure our lady readers [that it is] no connection of the monster in Frankenstein"—apparently oblivious to its own otherwise explicit association of one monster with another.[31]

However, musical objects were not confined to this assemblage of resonant ciphers for modernity (the clock, the piano, the bionic arm). Particularly decorative pianos appeared in Class XXVI ("Furniture, Upholstery, Paper Hangings, etc."), among them a controversial "Gothic Piano" placed in Augustus Pugin's equally controversial Mediaeval Court. The response of a critic writing for *Newton's London Journal* was curt but revealing—"We have already stated an objection to this kind of decoration"[32]—in ways that clarify the links between long-entrenched objections to ornamentation in virtuoso performance and negative responses to the construction of musical instruments. Nor was the idea that ornament (of any sort) would be detrimental to an instrument's musical effect restricted to the pianos on display. Tallis's *History and Description of the Crystal Palace* complained about instrument manufacturers' use of "elaboration, in order to effect a very simple object." Concerned above all by various innovations in horns and flutes, Tallis went on to insist: "Nothing injures tone more than a superabundance of mechanism."[33]

Yet it was precisely the internal mechanisms of musical instruments that were most widely spread and most difficult to classify at the Exhibition, with objects sorted according to the principal type of material or manufacture produced by each exhibitor. Wire, hinges and locks for pianos were scattered across the "General Hardware" section, along with plates for music printing; drum heads were included with "Shoes and Leather"; bell ropes and decorative fretwork for pianos appeared as "Manufactures from Animal and Vegetable Substances," while "specimens of lithographic music printing" and "ornamental printed music" were featured as the sole musical items in the Fine Arts Court. Entirely symptomatic of the status of these exhibits is the fact that nowhere is it recorded which pieces of music were offered as printed specimens. These instrumental components and associated technologies were displayed as just that: material artifacts largely divorced from consideration of their use in musical performance.

ON HEARING THINGS; OR,
GHOSTS IN THE MACHINE

As represented by this series of classified, staged objects, then, music seems to have been manifested in a striking multiplicity of ways within the Exhibition. Indeed, the sheer variety of these manifestations in the Crystal Palace suggests that music was quite literally out of place in this most overdetermined of Victorian taxonomies. Nevertheless, the various guidebooks, personal commentaries, and newspaper reports about the Exhibition— and perhaps above all the catalogues in which its contents were repeatedly listed—make the physical presence of musical objects abundantly clear. Individual musical exhibits are singled out for brief praise (or criticism) with regularity: most often the organs, but also other instruments gathered in the main British exhibit, and occasionally those placed in foreign sections elsewhere in the building. Yet none of these critical appraisals gives any sense that their authors were aware of the broader spread of component parts of musical instruments; none records anxiety about the presence of musical instruments in an exhibition of Works of Industry of All Nations. Indeed, the closest one comes to locating that sort of response—something reflecting explicitly on the status of musical instruments at the Exhibition—is Tallis's observation that the fact that time improves the tone of string instruments "gives to this department of the manufacture of musical instruments a color of antiquarianism (so to say), which possibly removed it beyond the world of contemporary enterprise represented in Hyde-Park."[34]

Tallis's image, of the musical past threatening to encroach on a celebration of the industrial present, stands out among the surfeit of passing comments on individual instruments. But even this provocative response might be largely countered (if not entirely dispelled) by the fact that one of the medals awarded in the musical instrument category was presented to French violin maker J. B. Vuillaume "for new modes of making violins, in such a manner that they are matured and perfected immediately on the completion of the manufacture, thus avoiding the necessity of keeping them for a considerable period to develop their excellences."[35] In the technologically advanced present of 1851, it seems that even the passing of time could be artificially manufactured. And the faint whiff of "antiquarianism" detected by Tallis in the musical instrument display was more symbolic than actual, since all objects in the Exhibition had to have been manufactured fewer than three years previously. As objects—however carefully or bizarrely ordered—these musical instruments thus remain stubbornly silent in the face of twenty-first-century interrogation.

What is ironic in this context is that these instruments were evidently played. In addition to their presence as material objects, to be admired for their qualities as artifacts, or as evidence of "progress" made in manufacturing techniques and design, the musical instruments on display at the Exhibition were experienced as a sounding means to another, much less obviously "industrial" end. Percy Scholes's survey of British musical culture between 1844 and 1944 as recorded in the *Musical Times* makes brief mentions of recitals by J. T. Cooper on Willis's organ, the organ music of Hesse played (and apparently made popular) on "various instruments," the "extempore fugues" on the organ built by Gray & Davison, and over forty recitals given by the pianist A. J. Hipkins on Broadwood's pianos.[36] Elsewhere, and despite the widespread absence of any explicit comment on the musical instrument displays as a whole, reports of visits to the Exhibition mention musical sounds as well as industrial noise: Martin Cawood's overwhelming sensory experience, with which this chapter began, was a cacophony of "pealing strains" and "brilliant tones." However, apart from a complaint (again from Tallis) about the ineptitude of the musicians demonstrating Sax's new instruments and the recitals mentioned in the *Musical Times* (and listed by Scholes), virtually no trace remains of which instruments were played, how often, and by whom. Given the absence of official mention in the Royal Commission's minutes—which discuss in detail the timings and practical arrangements made for the demonstration of the large machinery on display—we must assume that decisions concerning the demonstration of musical instruments were left to individual exhibitors. And although the explanation seems strange in the context of an event as minutely overseen as this one, an alternative is difficult to find. Once again, traces of past musical experience remain mute.

Such encounters with historical objects—and noisy, musical ones at that—which have somehow lost the power of communication in our musicological present, must, if nothing else, call for a change of approach. We might take as our cue an account written by one of the Exhibition's more distinguished musical visitors (and proud jury member), Hector Berlioz, who reported on his experiences in London in his regular *Journal des débats* column. Berlioz describes one occasion on which, unable to sleep after a particularly overwhelming concert in St. Paul's Cathedral, he decided to go to the Crystal Palace before it opened for the day. At 7:00 am, with the building deserted, he was deeply impressed by

> the vast solitude, the silence, the soft light falling from the transparent roof, all the stilled fountains, the silent organ, the motionless trees, and the well-blended display of goods brought there from every corner of the earth by a hundred rival nations. Those ingenious works, the products of peace, those instruments of destruction, reminiscent of war, all those fomenters of movement and noise, seemed then to be talking to each other mysteriously in man's absence, in that unknown language which one hears with the mind's ear.[37]

What is suggestive here is not simply that Berlioz marks the absence rather than presence of sound (musical or otherwise) but, more importantly, that he imagines in such absence an alternative form of communication: sounds emanating from the objects gathered in the Crystal Palace and audible in "the mind's ear." Invoking such apparently loquacious things might gesture in one sense toward the discourse surrounding material cultures and thingliness currently in favor in certain humanities circles; it would certainly be possible—perhaps also productive—to place a discussion of musical ontology ca. 1851 in dialogue with such ideas. In what remains of this chapter, however, I want to do something else, and switch my focus from asking what was on display in the Crystal Palace (those now-silent artifacts) to tracing the experience of the Exhibition visitor: a listener equipped with a mind's ear.

MATERIALIZING THE MUSICAL WORK

A week after the Great Exhibition opened to the public, the long bulletin from Hyde Park that had become a daily fixture in the *Times* was already showing signs of object fatigue. A perfect antidote, the writer suggested, would be found in musical performance:

> the overtaxed sight wishes and longs for relief in that great palace of wonders. The longer one stays and the oftener one visits the building, the more irresistible does the craving for music become. Everybody feels and expresses this want, and the occasional half-notes of an organ, or the faint

tinkling sounds of a piano, as they fall upon the ear, only aggravate the
general desire. That vast interior leaves ample scope for, and is suggestive
of, action in some shape or other. Nor can the public be left entirely to
the pleasures of meditation over inanimate forms and substances, however attractive.[38]

The hypothetical visitor to the Crystal Palace invoked here was both a de-
siring subject—one capable of craving musical "relief"—and an object in
need of animation. The "action" called for was not so much on the part of
the visitor herself, but rather something to be provided externally, as a more
potent form of stimulation than that offered by the mute "inanimate forms
and substances" on display for visual consumption. This visitor was, moreover, sensitized to musical sound to precisely the extent that her sight was
"overtaxed," and her powers of "meditation" weakened. After a week of sustained consumption, that is, the visually empowered exhibitionary subject
explored at length by Tony Bennett had become all ears: a listener as passive
as any modern (even modern*ist*), disciplined mode of musical attentiveness
might demand.[39]

As described by the *Times*, then, the army of objects on display van-
quished the consumer through sheer excess of information and weight of
numbers. As Charles Dickens (no fan of the Exhibition) complained, "I don't
say 'there's nothing in it'—there's too much. . . . So many things bewilder
me."[40] The impression here is of something approaching an overdose of
materiality. Such, moreover, is the power of a certain type of "thing": the
commodity whose staging and apparent celebration has so frequently been
thought to lie behind the Exhibition's claim to cultural modernity. Yet on
the basis of the *Times* report just mentioned, we might dispute Marx's famous claim in his *Grundrisse* that "consumption completes the act of production by giving the finishing touch to the product as such, by dissolving
the latter, by breaking up its independent material form."[41] On the contrary,
the ordered scenes of consumption characteristic of the Great Exhibition did
not so much dissolve the product as they did the consumer herself. There is,
after all, little trace of an individual experience in the *Times*'s description:
"everybody" is imagined to feel the same desire; and this corporate emotion
is intensified by the music heard (almost) always at a distance. Such distant
sounds not only prevent any degree of aural absorption; they also, by extension, seem to recast the Crystal Palace as a dispersive, "centrifugal" auditory
space, in contrast to what Richards has identified as the "centripetal" properties underpinning its economy of visual display (with the commodity once
again as its "centre and axis").[42] Any temptations toward aural attentiveness

at the Crystal Palace, that is, required the extension of the ear into the building's remotest corners—in search of sounds beyond the reach of even the most advanced ear trumpets on show among Class X's surgical implements.

In such surroundings it was, in the end, musical performers themselves whose presence at the Great Exhibition was most prone to instability, even dematerialization, both in 1851 and in the years since. The opening ceremony on May 1 featured no new musical commission, no high-profile solo performance, and no large-scale works. It did, however, involve considerable musical forces. In his program for the ceremony, the musical superintendent Sir George Smart ("Organist and Composer to Her Majesty's Chapel Royal") listed no fewer than 783 participating musicians, ranging from star soloists on loan from London's opera houses to a brace of eminent organists to three-figure-strong battalions of unnamed chorus members.[43] The repertoire essayed by this vast group was, perhaps inevitably, predictable and unimaginative: trumpet flourishes to punctuate proceedings; two mass renditions of the National Anthem at the start and close; a triumphant "Hallelujah Chorus" following a prayer led by the Archbishop of Canterbury; unspecified contributions by the bands of the Coldstream and Scotch Fusilier Guards. These sonic explosions were reported (and have since been discussed) largely for their statistical interest: as a high-profile instance of the Victorian "monster concert" phenomenon. But what is immediately striking is that the massed musical forces seem to have been barely heard in the Crystal Palace. According to a reporter of the greatest conceivable eminence (Queen Victoria, in her diary), the "200 instruments and 600 voices . . . seemed nothing."[44] Enclosed within the great glass display case of the Exhibition, performing musicians were not so much elevated to the state of an exhibit as dissolved in the moment of their consumption: here, surely, is a musical instance of Marx's much-repeated dictum that "all that is solid melts into air."[45]

In the capitalist cathedral of the Crystal Palace and amid its clamorous celebration of material things—in gleaming array, stretching as far as the eye could see—music vanished and became inaudible. Peculiarly lacking in substance, it left hundreds of mute bodies in its wake. Standing metonymically for Handel's great musical work (perhaps even for "the great musical work" as the emerging foundation of elite musical culture), the "Hallelujah Chorus" proved a frail object, having in performance none of the solidity and permanence of the commodities on display. Here we are confronted at last with our own, widely used sense of the phrase "musical object," as a shorthand—perhaps even a euphemism—for the part-abstract concept, part-resonant phenomenon once (briefly) known as The Music Itself and, before that, simply called "music." Yet in its fragile, sounding form, music's place in this overwhelmingly object-oriented context seems a far cry from its episte-mological state as sketched by Lydia Goehr:

> As it entered the world of fine arts, music had to find a plastic or equiva-lent commodity, a valuable and permanently existing product, that could be treated in the same way as the objects of the already respectable fine arts. . . . The object was called "the work."[46]

In the 1851 Great Exhibition, music was indeed present as "a plastic or equivalent commodity, a valuable and permanently existing product"; but such commodities, and such products, were largely mute. No one denied the plasticity—the blunt materiality—of the lengths of piano wire, the drum heads, bell ropes, faux-antique violins, transposing pianos, or new, improved flutes. These were musical objects to be sorted, marveled at, and perhaps even (in due course) possessed; they were "Works of Industry," signs of prog-ress. Yet some of these musical objects were also machines in their own right. They were musical instruments displayed as mechanisms for the generation of musical sound: for the reproduction of an altogether less solid musical object. The musical displays at the Exhibition, in other words, once again offered two quite different perspectives on "music": one that demonstrated, triumphantly, the progress made in technological innovation; another that deferred implicitly to the universalism of "great works" and an imaginary museum increasingly at the heart of mid-century elite culture. But in those ephemeral musical performances at the Crystal Palace (whether at the open-ing ceremony or in demonstrations, seemingly ever-distant, of instruments on display), the separation of these types suddenly collapsed. In the echoing dream house of the Exhibition, music was presented not merely as collec-tion of polished things in glass cases. Rather it was actualized as an object that dissolved the instant it was produced, before its promise of materiality had been fulfilled. The musical "work" may have originated as the art form's entrance ticket to the cultural pantheon, in other words, but the Great Exhi-bition's industrial pageant cast it in an altogether more problematic light: as a shining example of the commodity form.

NOTES

1. "The Great Exhibition," *Leeds Mercury*, May 24, 1851.
2. Ibid.
3. Lydia Goehr, *The Imaginary Museum of Musical Works*, 2nd ed. (Oxford: Oxford University Press, 2007).

4. Jericho, "Exhibition of the Industry of All Nations," *Eliza Cook's Journal*, May 1850: 217.

5. Thomas Carlyle, "Signs of the Times," *Edinburgh Review* 98 (1829): 317.

6. This Birmingham Exhibition led in due course to the construction in 1850 of Bingley Hall, the country's first purpose-built industrial exhibition hall. More thought provoking than its claim to novelty, however, is the fact that the building was apparently constructed using girders left over from the construction of Euston railway station in London, which were transported on the London and Birmingham Railway (which had opened in 1838). This gestures toward an unexpectedly literal instantiation of Walter Benjamin's famous association between nineteenth-century railway stations and exhibition buildings as iron structures sharing "transitory purposes"; see Benjamin, *The Arcades Project*, ed. Rolf Tiedemann, trans. Howard Eiland and Kevin McLaughlin (Cambridge MA: Belknap Press, 1999), 154. For a brief but useful overview of earlier exhibitions mounted in Britain and elsewhere, see Richard D. Altick, *The Shows of London* (Cambridge MA: Belknap Press, 1978), 455–56.

7. The crucial meeting during which Albert proposed such an exhibition was held at Buckingham Palace on June 30, 1849; the attendees were Thomas Cubitt, Henry Cole, Francis Fuller, John Scott Russell, and Prince Albert himself. The minutes note that "His Royal Highness communicated his views regarded the formation of a Great Collection of Works of Industry and Art in London in 1851, for the purposes of Exhibition, and of competition and encouragement"; "1851 Exhibition: Correspondence and Papers: 1849," A/1849, Archive of the Royal Commission for the Exhibition of 1851, Imperial College London.

8. *Official Catalogue of the Great Exhibition of the Works of Industry of All Nations, 1851* (London: Spicer Brothers, 1851), 5.

9. *Official Catalogue of the Great Exhibition*, 5.

10. Jericho, "Exhibition of the Industry of All Nations," 217.

11. Ibid.

12. "The Great Industrial Exhibition," *Athenaeum*, October 18, 1851: 1094.

13. James Buzard, Joseph W. Childers, and Eileen Gillooly, eds., *Victorian Prism: Refractions of the Crystal Palace* (Charlottesville: University of Virginia Press, 2007), 1. For other modern accounts with comparable historiographical ambitions for the Exhibition, see, e.g., Isobel Armstrong, *Glassworlds: Glass Culture and the Imagination 1830–1880* (Oxford: Oxford University Press, 2008); Carla Yanni, ed., *Nature's Museums: Victorian Science and the Architecture of Display* (London: Athlone Press, 1999); Paul Young, *Globalization and the Great Exhibition: The Victorian New World Order* (Basingstoke, UK: Palgrave Macmillan, 2009).

14. In marked contrast, for the next exhibition held in London, in 1862, a quartet of eminent composers were invited to contribute new works to a large-scale opening ceremony: William Sterndale Bennett (representing England), Auber (representing France), Meyerbeer (representing Prussia), and Rossini (representing Italy). Rossini refused the invitation and was promptly replaced by Verdi, who—uncharacteristically—accepted, only to be told that his *Inno delle Nazioni* could not be performed as planned. For more on the saga that unfolded in relation to these commissions, see Roberta Montemorra Marvin, introduction to *The Works of Giuseppe Verdi, Series IV: Hymns*, ed. Montemorra Marvin (Chicago: University of Chicago Press, 2007), xi–xxii.

15. Michael Musgrave, *The Musical Life of the Crystal Palace* (Cambridge: Cambridge University Press, 1995), 8, 9.

16. Michel Foucault, *The Order of Things: An Archaeology of the Human Sciences* (London: Routledge, 2002); originally published as *Les Mots et les Choses (Une archéologie des sciences humaines)* (Paris: Gallimard, 1966).

17. See Foucault, *The Order of Things*, 140–44.

18. Harriet Ritvo, *The Platypus and the Mermaid* (Cambridge, MA: Harvard University Press, 1997), 20–21.

19. Binominal nomenclature was first applied consistently to animals in the tenth edition of *Systema naturae* (1758); see Ernst Mayr, *Principles of Systematic Zoology* (New York: McGraw-Hill, 1969), 58.

20. Mayr, *Principles of Systematic Zoology*, 58.

21. Ibid., 59.

22. Andrew H. Miller makes a similar point when he connects the Great Exhibition's staging of progress to the topos (familiar from Wolfgang Schivelbusch's work in particular) of the railway's "annihilation" of time and space; in Miller's words, the Exhibition "concentrated space so as to make time—understood as the relative historical progress of the various nations—more easily perceived. . . . [It] brought objects and people from across the world to a single point, where their rates of development could be measured and compared"; Miller, *Novels Behind Glass: Commodity Culture and Victorian Narrative* (Cambridge: Cambridge University Press, 1995), 55.

23. Tony Bennett, "The Exhibitionary Complex," in *Culture/Power/History: A Reader in Contemporary Social Theory*, ed. Nicholas B. Dirks, Geoff Eley, and Sherry B. Ortner (Princeton, NJ: Princeton University Press, 1994), 140.

24. For more on Playfair, see Steve Edwards, "The Accumulation of Knowledge, or, William Whewell's Eye," in *The Great Exhibition of 1851*, ed. Louise Purbrick (Manchester, UK: Manchester University Press, 2001), 35. As a result of the building's physical division into "foreign" and "British" parts, in which only the latter were classified according to the Royal Commission's taxonomy, and since I am concerned with what we might uncover about the musical culture of mid-nineteenth-century London in particular, this chapter necessarily addresses only those musical objects exhibited by British manufacturers. It goes almost without saying that still more work remains to be done on the display of non-European instruments at the Exhibition; the one relatively comprehensive account (albeit one in catalogue form) of musical instruments on show across all parts of the Exhibition is Peter and Ann MacTaggart, *Musical Instruments at the 1851 Exhibition* (Welwyn: Mac and Me, 1986).

25. Thomas Richards, *The Commodity Culture of Victorian England: Advertising and Spectacle, 1851–1914* (London: Verso, 1991), 17.

26. Quoted in Andrea Hibbard, "Distracting Impressions and Rational Recreation at the Great Exhibition," in Buzard et al., *Victorian Prism*, 151.

27. Richards, *The Commodity Culture of Victorian England*, 17.

28. Minutes of the Fourth Meeting of the Royal Commission (January 31, 1850), 8–9, RC/8/A/2/1, Archive of the Royal Commission for the Exhibition of 1851, Imperial College London.

29. Ibid., 3.

30. The presence of various innovations in self-playing instruments within the musical displays offers still more food for thought in this regard.

31. *A Guide to the Great Exhibition* (London: George Routledge, n.d.), 193.

32. Quoted in MacTaggart, *Musical Instruments at the 1851 Exhibition*, 34.

33. John Tallis, *History and Description of the Crystal Palace and the Exhibition of the World's Industry in 1851.* 3 vols. (London: John Tallis, 1851), 1:119.

34. Ibid., 3:59.

35. John Timbs, *The Year-Book of Facts in The Great Exhibition of 1851* (London: David Bogue, 1851), 306.

36. Percy Scholes, *The Mirror of Music, 1844–1944* (London: Novello, 1947), 853, 595, 601 and 854, 776.

37. Hector Berlioz, "Twenty-first Evening," in *Evenings in the Orchestra*, trans. C. R. Fortescue (1852; London: Penguin, 1963), 211–12. Berlioz's account originally appeared as a letter to the editor (dated June 9) in the *feuilleton* of the *Journal des débats* (June 20, 1851): 1–2.

38. "The Great Exhibition," *Times* (London), May 7, 1851.

39. According to Bennett, exhibitionary institutions such as the Crystal Palace "sought not to map the social body in order to know the populace by rendering it visible to power—the power to command and arrange things and bodies for public display—they sought to allow the people, *en masse* rather than individually, to know rather than be known, to become the subjects rather than the objects of knowledge"; Bennett, "The Exhibitionary Complex," 126.

40. Letter to Mrs Watson (July 11, 1851), *The Letters of Charles Dickens*, Pilgrim ed., ed. Graham Storey, Kathleen Tillotson, and Nina Burgis. 12 vols. (Oxford: Clarendon Press, 1988), 6:428, cited in Juliet John, *Dickens and Mass Culture* (Oxford: Oxford University Press, 2010), 63.

41. *Marx's "Grundrisse,"* ed. and trans. David McLellan (London: Macmillan, 1971), 25.

42. Richards draws a distinction between the "centrifugal" space of the advertisement, which he figures as directing attention away from the center of representation, and the Crystal Palace's "centripetal space of representation that took the commodity as its centre and axis"; Richards, *The Commodity Culture of Victorian England*, 52–53.

43. "Exhibition of 1851. Programme of the Musical Performances, May 1st, 1851." Papers of Sir George Thomas Smart, vol. 7: MSS/Additional/41777, 280ff, British Library, London.

44. "May 1, 1851," Queen Victoria's Journals [Princess Beatrice's copies], vol. 31, p. 213, accessed October 2, 2013, http://www.queenvictoriasjournals.org/home.do, RA VIC/MAIN/QVJ (W) May 1, 1851 (Princess Beatrice's copies), retrieved October 2, 2013.

45. The most famous use of this phrase is Marshall Berman's in *All That Is Solid Melts into Air: the Experience of Modernity* (London: Verso, 1983). It is no coincidence that Berman includes a brief discussion of the Crystal Palace as "one of the most haunting and compelling of modern dreams" (236) and quotes an account of the Palace's interior that he reads as exemplifying the tendency toward decomposition observed by Marx: "If we let our gaze travel downward it encounters the blue-painted lattice girders. At first these occur only at wide intervals; then they range closer and closer together until they are interrupted by a dazzling band of light—the transept—which dissolves into a distant background where all materiality is blended into the atmosphere" (239).

46. Goehr, *The Imaginary Museum of Musical Works*, 173–74.

ACKNOWLEDGMENTS

This volume was the outcome of a series of study days for Music in London, 1800–1851, a research initiative based at King's College London and funded by the European Research Council. We are grateful to our interlocutors at this early stage, including James Chandler, Eleanor Cloutier, James Grande, Katherine Hambridge, Jonathan Hicks, Oskar Cox Jensen, Arman Schwartz, Benjamin Walton, Heather Wiebe, and Alison Winter. Profuse thanks are also owed to the two anonymous readers of the manuscript, who are among the most insightful and generous unnamed humanists we have encountered. The volume became a book under the guidance of Marta Tonegutti and Evan White at the University of Chicago Press. Above all, the editors would like to thank Roger Parker, who oversaw this project from its beginnings. As ever, Roger discovered our various writings "minutely and multitudinously scratched in all directions; . . . and lo! the scratches [seemed] to arrange themselves in a fine series of concentric circles" (George Eliot, *Middlemarch*).

CONTRIBUTORS

JAMES Q. DAVIES (PhD Cambridge University, 2005) is associate professor of music at the University of California, Berkeley. He is author of *Romantic Anatomies of Performance* (University of California Press, 2014). Articles and essays appear in *19th-Century Music* (2003), *Cambridge Opera Journal* (2005), *Royal Musical Association* (2006), *Opera Quarterly* (2006; 2012), *Keyboard Perspectives* (2010), *Representations* (2015), and the *Journal of the American Musicological Society* (2015).

MELISSA DICKSON is a postdoctoral research assistant at St Anne's College, Oxford, and is currently working on explorations of the body's physiological and psychological responses to sound and music in the nineteenth century.

EMILY I. DOLAN is the Gardner Cowles Associate Professor of Music at Harvard University. Her work focuses on instrumentality and the intersections of music, science, and technology. She is the author of *The Orchestral Revolution: Haydn and the Technologies of Timbre* (Cambridge University Press, 2013).

SARAH HIBBERD is associate professor of music at the University of Nottingham. Her research focuses on opera and musical culture in Paris and London. Her publications include *French Grand Opera and the Historical Imagination* (Cambridge University Press, 2009) and a special issue of *19th-Century Music* (Fall 2015) devoted to music and science in London and Paris, for which she was guest editor.

MYLES W. JACKSON is the Albert Gallatin Research Excellence Professor of the History of Science at NYU–Gallatin, professor of history in the College of Arts and Science at NYU, and director of the Program in Science and Society at NYU. He has authored three books: *Spectrum of Belief: Joseph von Fraunhofer and the Craft of Precision Optics* (MIT Press, 2000), *Harmonious Triads: Physicists, Musicians, and Instrument Makers in Nineteenth-Century Germany* (MIT Press, 2006), and *The Genealogy of a Gene: Patents, HIV, and Race* (MIT Press, 2015). He is the coeditor, with Alexandra Hui and Julia Kursell, of *Music, Sound, and the Laboratory from 1750 to 1980* (University of Chicago Press, 2013) and guest editor of *Gene Patenting,* a special issue of *Perspectives on Science* (2015).

ELLEN LOCKHART (PhD Cornell University, 2012) is assistant professor of musicology at the University of Toronto. Her monograph *Animation, Plasticity, and Music in Italy, 1770–1830* is forthcoming from the University of California Press, and her articles have appeared in *Eighteenth-Century Music* and the *Cambridge Opera Journal*. Her critical edition of Puccini's *La fanciulla del West* will be published by Ricordi and was performed at La Scala in 2015 under Riccardo Chailly; she is also coediting, with David Rosen, a critical edition of the opera's original *mise-en-scène*. She has recently become reviews editor for the *Cambridge Opera Journal*.

DEIRDRE LOUGHRIDGE (PhD University of Pennsylvania) is a musicologist at the University of California, Berkeley. She has held fellowships from the Mellon Foundation and the American Council of Learned Societies. Her articles have been published in *Eighteenth-Century Music, Journal of Musicology,* and *Journal of the Royal Musical Association*. Her book, *Haydn's Sunrise, Beethoven's Shadow: Audiovisual Culture and the Emergence of Musical Romanticism,* was published by University of Chicago Press in July 2016.

GAVIN WILLIAMS is a research fellow at Jesus College Cambridge; his research focuses on the entanglement of music with media technologies during the nineteenth and twentieth centuries, and he is currently editing a volume about the sounds of the Crimean War.

FLORA WILLSON is a British Academy postdoctoral fellow in the Music Department at King's College London, having previously held a junior research fellowship at King's College Cambridge. She has published in journals including *Cambridge Opera Journal, Opera Quarterly,* and *19th-Century Music,* as well as in the *Cambridge Verdi Encyclopaedia* (Cambridge University Press, 2013) and the *Oxford Handbook of the Operatic Canon* (Oxford University Press, forthcoming). She is currently working on a book about the international networks underpinning late nineteenth-century operatic culture. Flora is also the editor of a new critical edition of Donizetti's 1840 grand opera *Les Martyrs* (Ricordi, 2015), which premiered at London's Royal Festival Hall in November 2014.

INDEX

Page numbers in italic refer to figures.

Abel, Carl Friedrich, 9, 14
Abernethy, John, 176–77, 196–97
Académie des Sciences, 103, 115
Academy of Ancient Music, 8, 14, 225
Ackermann's Repository, 133–34
acoucryptophone. *See* Wheatstone, Charles: enchanted lyre
acoustics, 2, 4, 5, 21–22, 27–28, 89–90, 105–15, 120–21, 125–42, 155–59
Adorno, Theodor, 78, 96
aeolian harp, 60, 93–95, 139–40, 178, 189–90, 193, 197
aeolina, 117–19
aeolodicon, 117
Agnew, Vanessa, 40
Albert, Prince Consort, 229, 234
Alhambra Palace. *See* Royal Panopticon of Science and Art
Alison, Archibald, 191
Allen, William, 64
Altick, Richard, 6, 136
Antarctic, the, 163
Apollonicon, 6
Arago, François, 115
Arden, John, 4, 54
Argyll Rooms, 13–16
Arne, Michael, 5, 63–64
Arnold, Joseph, 152
artificial intelligence, 23, 203–11, 219–20
Asiatic Society, 160
astronomy, 9, 20, 30–33, 41–42, 47–72, 160, 206–7
Athenaeum, The, 13, 181, 230

Auber, Daniel, 6
automata, 5–6, 23, 101–5, 117, 161, 209–10; automated orchestra, 6, 101–2
Ayrton, William, 13

Babbage, Charles, 4, 12, 23, 103–4, 140, 203–20
Bach, Carl Philipp Emanuel, 20, 35
Bach, Johann Christoph, 9, 14
Bach, Johann Sebastian, 17, 168
Bacon, Francis, 18
Baddeley, Sophia, 5
Banks, Sir Joseph, 12, 20, 40
Barker, Henry, 104. *See also* panorama
barrel organ, 57, 88, 212–19
Bass, Michael, 211, 213, 216–17, 220
Bassi, Laura, 34–35
bassoon, 22, 38, 101, 108
baton conductor, 2, 14–15
Beethoven, Ludwig van, 6, 14, 15, 78, 79, 82
Bell, Alexander Graham, 22, 118, 161
Bell, Charles, 178
Bellini, Vincenzo, 9, 10, 159
bells, 23, 38, 210–11, 219–20, 222
Bennati, Francesco, 159
Bennett, Tony, 233, 240
Bennett, William Sterndale, 79, 96, 243
Berlin, 39
Berlioz, Hector, 9, 17, 19, 78–79, 165, 238–39
Biot, Jean-Baptiste, 115, 158
Birchall & Co., 8
Birmingham, 65–66, 229, 243
Birmingham Exhibition of Manufactures and Art, 229

Blake, William, 32–33
Blasius, Leslie David, 191, 201
Bochsa, Nicolas-Charles, 13
Boia, Michael, 19
Bombay, 146
Boosey & Co., 8
Boscovich, Roger, 20, 33–34
bows and bowed instruments, 3, 11, 19–22, 24,
    41, 47, 57–60, 68, 89–90, 106–9, 116, 130–32,
    156, 189, 202, 237, 242. *See also* celestina;
    coelestinette; violin; violoncello
Boyle, Robert, 36
Braid, James, 14
brass instruments and bands, 10, 205, 212–13.
    See also trombone; trumpet
Brewster, David, 104, 108, 131. *See also*
    kaleidoscope
British Association for the Advancement of
    Science, 117, 158
British East India Company, 150, 152–54, 170
British Institution, 12
Brontë, Charlotte, *Jane Eyre*, 140–42
Brunel, Isambard, 145. *See also* Great Eastern
Brussels, 106
Bülow, Hans von, 79
Bunsen, Christian von, 160–61
Burghersh, John Fane, Lord, 13
Burney, Charles, 2, 4, 19–20, 23, 27–43, 68, 75,
    191; *General History of Music*, 2, 19, 24,
    39–43, 191; *An Essay towards a History of
    the Principal Comets that have Appeared
    since the Year 1742*, 30; *Astronomy*, 31–33;
    *Verses in praise of Haydn's arrival in
    England*, 32
Burney, Charles Rousseau, 5, 63–64
Burney, Edward Francis, 47–48, 56, 63
Burney, Elizabeth, 31
Burney, Fanny, 31, 33
Burney, Susannah, 82
Burwick, Frederick, 93

calculating machine. *See* difference engine
Callcott, John Wall, 40
Cambridge University, 101, 103, 159, 167, 203–4
Canada, 165
Carlyle, Thomas, 229
Cassini, Jacques, 30–31
Catalani, Angelica, 8, 159, 192
Cavendish Laboratory (Cambridge University),
    101
Cawood, Martin, 227–28, 238
celestina, 3, 20, 47–60, 65–70

cello. *See* violoncello
Chabert, Ivan, 15
Chandler, James, 188
Chappell & Co., 8
Chassebœuf, Constantin-François de, 160
Cherubini, Luigi, 19
Cheshire, 51
China, 41–42, 165
Chinese music, 41, 153, 158
Chladni, Ernst Florens Friedrich, 106–7, 115,
    121–22, 129–31; Chladni figures, 106–8,
    129–31, *130*, 156–57
Chopin, Frédéric, 9
Chorley, Henry, 162
Cimarosa, Domenico, 192
clavichord, 20, 35
clavicylinder, 106
Clement, Joseph, 208–9
Clementi, Muzio, 9; Clementi & Co., 8
Coburg Theatre, 23, 184–85, 200
Cocks & Co., 8
coelestinette, 3, 59–60
Coleridge, Samuel Taylor, 33, 93–95, 193
Colò, Angelo, 181
Conan Doyle, Arthur, 140
concert halls and concert-hall listening, 1, 7,
    14–19, 181, 220
concertina, 3, 22, 118–20, *119*, 145–50, 161–71
concerts, 2, 4, 9, 10, 21, 82–85, 102, 125–28, 162,
    239, 241
conductor. *See* baton conductor
Conduit Street, Hanover Square, 47, 49, 101
Cook, James, 40
Cooke, T. P., 185, *186*, 189
Corder, Frederick, 77–79, 96
Court Magazine, 9
Covent Garden, 15, 63, 66, 103, 184
Cox, James, 103
Cramer, Johann Baptist, 9; Cramer & Beale's
    Music Warehouse, 8
Crary, Jonathan, 37, 197, 224
Crawfurd, John, 152–53, 155
Cross, Edward, 9
Crotch, William, 5, 7, 12, 153–55, 166, 201
Crystal Palace, 2, 10, 23–24, 229, 231, 233–37,
    239–42. *See also* Great Exhibition of the
    Works of Industry of All Nations
Curwen, John, 165, 170
cyclorama. *See* Royal Cyclorama

Darwin, Charles, 233
Darwin, Erasmus, 32–33, 178–80

Daston, Lorraine, 227
Davy, Humphry, 7, 18, 176, 182–83, 199, 207
Debussy, Claude, 78
Desaguliers, John Theophilius, 54–55
Dickens, Charles, 11, 162, 240
Dictionary of Music and Musicians (Sir
 George Grove), 78–79
difference engine, 3, 23, 203–20
Disney, Walt, 78, 96
Dorset Street, Marylebone, 23, 205, 208, 211–15,
 219
Dragonetti, Domenico, 9
Drayton Station (London), 146
Dublin, 25, 84, 117–18, 158
Dublin Evening Post, 120
During, Simon, 135
Dussek, Jan Ladislav, 9, 86

ear trumpet, 236, 241
Eastcott, Richard (*Sketches of the Origin,
 Progress, and Effects of Music*), 191
Eastern Telegraph Company, 146, *151*
echo, 27–30, *29*, 170, 242
Edinburgh, 86
Egypt, 146
Egyptian Hall, 19, 161
eidophusikon, 3, 5–6, 14, 20, 61–66, *63*, 72
eidouranion, 20, 47–72, *48*
electricity and electrical science, 52, 61, 137–
 42, 146, 176–82, 192–95
electric telegraph. *See* telegraphs and telegraphy
Eliot, George, *Middlemarch*, 137–38
Eliza Cook's Journal, 229–30
Ellis, Alexander J., 155, 160–70
enchanted lyre. *See* Wheatstone, Charles
English Opera House, 23, 47–48, 104, 175, 184–
 85, 193, 200
eolian harp. *See* aeolian harp
Ethnological Society of London, 160
Euler, Leonard, 98, 158, 165, 168
euphone, 106
European Magazine, and London Review, 51,
 60–62, 64
Exeter Hall, 16–18

Faber, Joseph, 161
Family Oracle of Health, 181
Fantasia (Disney), 78–79
Faraday, Michael, 7, 115–21, 132–33, 156
Farringdon Road, 212
Ferguson, James, 52, 54, 61
Fétis, François-Joseph, 9

Fitzwilliam Virginal Book, 79
flute, 11, 22, 38, 101, 108, 113, 188, 236
fortepiano. *See* pianos
Foucault, Michel, 232
Frankenstein; or the Man and the Monster!.
 See Milner, Henry
Frankenstein, or The Modern Prometheus. See
 Shelley, Mary
Freischütz, Der (Weber), 104
Fyfe, Aileen, 106

Gallery of Illustration, 5–6
Gallery of Practical Science, 6
galvanism, 181–83, 193
gamelan, 152–55, 166–67. See also *gendèr*
Gardiner, William, 94
Gauntlett, Henry, 15
Gawboy, Anna, 164–65
gendèr, 5, 22, 113–16, *114*, 154–60, 166, 170
George II, 11
George III, 41, 102
George IV, 10
Gillray, James, 47, 49, 56
Gloucester, 101
Goehr, Lydia, 7, 228, 242
Golden Square, 11, 67
Golinski, Jan, 55
Goulding & D'Almaine's Music Storehouse, 8
Grant, Robert Edward, 8
Great Eastern, 146–47, *150*
Great Exhibition of the Works of Industry of
 All Nations, 2, 23–24, 165, 227–42
Greatorex, Thomas, 14
Great Room, Spring Gardens, 5, 102–3
Great Western Railway, 146
Grieve, Thomas, 5–6
Grove, Sir George, 10, 78, 97
guimbarde. See Jew's harp

Haddock's Androides, 104
Hadley, Elaine, 184, 199
Haeckl, Anton, 117
Halley, Edmond, 31
Handel, George Frideric, 9, 69, 71, 168, 231,
 241
Hanover Square, 8, 47, 49, 60–61, 101; Hanover
 Square Concert Rooms, 13–14, 26, 84, 138
Harmonic Institution, 13–15
Harmonicon, 12, 28, 30
harmonium, 3, 168
harpsichords, 20, 47, 57–58, 63–64, 67–68, 86,
 103

Hartley, David, 178–80, 187, 191
Haskell, Yasmin, 32
Hauptmann, Moritz, 164–65
Havergal, Frances Ridley, 139–40
Hawkins, Sir John, 40
Haydn, Joseph, 9, 19, 32, 41, 69, 71, 82
Haymarket, the, 11, 60, 103–4, 138
Hays, J. N., 61
Helmholtz, Hermann von, 134–35, 159, 167–68
Herschel, John, 160
Herschel, William, 20, 31–32, 41–42, 69–70
Holborn Street, 212
Holland, 106
Hollis Street, 8
Holst, Gustav, 21
Hooke, Robert, 50
Hornbostel, Erich von, 165
Hughes, David Edward, 147. See also printing
    telegraph
Hullah, John, 17
Humboldt, Wilhelm von, 153
Hyde Park, 10, 18, 23, 229, 238–39. See also
    Crystal Palace; Great Exhibition of the
    Works of Industry of All Nations

Imperial Academy of St. Petersburg, 158
India, 141, 165
Inkster, Ian, 72
International Exhibition (London, 1862), 210
International Inventions Exhibition (South
    Kensington, 1885), 166
invisible girl. See Wheatstone, Charles
Islington Literary and Scientific Institution,
    15

Jacob, Margaret, 55
Jacquet-Droz, Henri-Louis, 103
Jakarta, 150
Jane Eyre (Brontë), 140–42
Japan, 165
Jefferson, Thomas, 47, 67–68
Jewin Street Independent Chapel, 170
Jew's harp, 114, 156, 158
John Napier Society, 203
Jullien, Louis, 9, 10, 13

kaleidophone, 3, 21, 108–11, 109, 131
kaleidoscope, 3, 108–9, 131
Kauer, Ferdinand, 86
Keats, John, 140
Kempelen, Wolfgang von, 104, 118, 158
Kennaway, James, 193

keys, keyboards, and keyboard instruments, 8,
    11, 14, 20–21, 24, 33, 35, 41, 47, 57–60, 63–68,
    73, 75, 79–86, 102–3, 108, 113, 117–20, 125,
    135, 145–50, 146–49, 159, 161–70, 169, 212,
    227, 231, 236–38, 240, 242. See also bas-
    soon; celestina; clavichord; clavicylinder;
    concertina; flute; harpsichords; organ and
    organ pipes; pianos; printing telegraph;
    symphonium; typewriter
King's College London, 8, 22, 120–21, 148, 155–
    56, 162; Wheatstone Collection, 146, 162
King's Theatre, 9, 11, 19, 79, 82, 138, 192
Kipling, Rudyard, 140
Kircher, Athanasius, 27–29
Klancher, Jon, 12
Koczwara, František, 86
Kornhaber, David, 62
Kratzenstein, Christian Gottlieb, 158

Lablache, Luigi, 9, 159
Lacey, Michael, 5–6
Landon, H. C. Robbins, 32
Langley, Leanne, 14
Laplace, Pierre Simon, 90
Lardner, Dionysius, 12
Lawrence, William, 176–78, 182–83, 196–97
Leicester, 84
Leicester Square, 17, 64, 104
Leppert, Richard, 22, 126
Lessing, Gotthold Ephraim, 94
Levere, Trevor, 93
Lightman, Bernard, 7, 106
Lind, Jenny, 9, 17
Linnaeus, Carl, 32, 232–33
Lissajous, Jules Antoine, 123, 131
Liszt, Franz, 9, 15, 77–79
Literary Gazette, 125, 129, 133
Literary Panorama, 181
London Lions, 70–72
London Magazine, 13, 175
London Missionary Society, 16
London Morning Post, 181, 184
London Musical Association, 166, 170
London Phrenological Association, 155
London University. See University of London
Lonsdale, Roger, 31–32
Loutherbourg, Philippe-Jacques de, 20–21, 61–
    67
Lussier, Mark, 32, 93
Lyceum. See English Opera House
Lyndhurst, John Copley, Baron, 218–19
lyrichord, 3, 57–58, 59

Madagascar, 165
Madame Tussaud's, 5
Maillardet, Henri, 103
Mainzer, Joseph, 17
Malus, Étienne, 111–12
Mälzel, Johann Nepomuk, 104
Manchester, 51–52
Manchester Square, 213–14
Martin, Benjamin, 54, 58–61
Martini, Padre (Giovanni Battista), 20, 40
Marx, Karl, 1–4, 24–25, 218, 240–45
Mathew, Nicholas, 96
Maupertuis, Pierre, 30–31
Mayr, Ernst, 233
McKeon, Michael, 38
McLandburgh, Florence, 140
Mecklenburg-Strelitz, Charlotte of, 102
Meisel, Martin, 24
melodrama, 23, 95, 177, 183–90
Mendelssohn, Felix, 7, 14–17, 21, 78, 168
Menke, Richard, 140–41
Méric-Lalande, Henriette, 159
Merlin, John Joseph, 103. See also Merlin's
    Mechanical Exhibition
Merlin's Mechanical Exhibition, 6, 23, 103–4
Mesmer, Franz Anton, 180–81
mesmerism, 12, 181
Messier, Charles, 31
Metropolitan Institution, 12
Metropolitan Police Service, 211
microphony, 2, 147, 157
Middlemarch (Eliot), 137–38
Milan, 27, 30, 33–34
Milner, Henry, 23, 177, 183–95
Molique, Bernhard, 162
Moody, Jane, 184, 190
Moore, Thomas, 136–37
Mori & Lavenu, 8
Morning Chronicle, 176, 205
Moulton, John Fletcher, 203–6
Mozart, Wolfgang Amadeus, 6, 14, 96
Müller, Johann, 37
Müller, Max, 160
Musgrave, Michael, 231
Musical Institute of London, 13
Musical Times, 17, 167–68, 238
Musical World, 15–17, 102, 133, 162
musicology, emergence and history of, 2, 12–
    20, 27–30, 33–43, 78–79, 96–97, 165–66, 228

Napoleon, 5, 106
Nash, John, 10–15

New Grove Dictionary of Music and Musi-
    cians, 77–79. See also Dictionary of Music
    and Musicians (Sir George Grove)
New Monthly Magazine, 9, 13, 134, 234–35
Newton, Isaac, and Newtonian science, 18, 34,
    37, 47, 50, 52, 53, 56, 60, 89–90, 181
Newton's London Journal, 236
New Zealand, 165
Nicholas Nickleby. See Dickens, Charles
Noble, William, 163
Novello, J. Alfred, 16–17

oboe, 41, 202
oedephone, 70, 102
Oettingen, Arthur von, 164
Old Price riots, 184
optics, 3, 34, 44, 52, 56, 61, 69, 89–90, 104, 108–
    11, 120–21, 131
organ and organ pipes, 5, 33–34, 41–42, 57, 67–
    68, 89, 133, 156, 165, 168, 170–72, 231, 236,
    239. See also barrel organ
Oriental Repository, 155, 170
orreries, 20, 35–36, 47–72, 55, 104. See also
    eidouranion
Ørsted, Hans Christian, 122, 130–31
Owen, Richard, 12, 160

Paddington Station (London), 146
Pall Mall, 11, 61, 102, 125, 145
panorama, 5, 62, 104
Pasta, Giuditta, 181
Peake, Richard Brinsley, 23, 175–77, 183–95
phenakistoscope, 3
Phillips, William, 94
phonetics, 159–66; International Phonetic
    Alphabet, 161
photograph, 3, 150
phthongometer, 156–61. See also speaking
    machine
pianos, 8, 11, 14, 21, 24, 57, 60, 79–86, 102, 108,
    118, 125, 127–28, 135, 145, 149, 159, 164–65,
    168, 170, 212, 227, 231, 236–38, 240, 242
Planché, James Robinson, 104
Playfair, Leon, 234
Plenius, Roger, 57–59
Polytechnic. See Royal Polytechnic
    Institution
Presumption; or, The Fate of Frankenstein.
    See Peake, Richard Brinsley
Priestley, Joseph, 60–61, 65, 182–83
printing telegraph, 145–46, 147
program music, 77–97

Prony, Gaspard de, 207–8
Purcell, Henry, 18

Quarterly Musical Magazine and Review, 8, 13

Raden Rana Dipura, 152–53, 155
Raffles, Sir Stamford, 150–56, 160
Rancière, Jacques, 215
reeds and reed instruments, 22, 114–19, 156–63. See also bassoon; oboe; speaking machine
Regent Street, 5–8, 10, 12–13, 18
Regondi, Giulio, 120, 162, 164
Reid, David Boswell, 15
Repton, Humphry, 10
resonance, 5, 21–22, 113–16, 156–59, 163–64, 168–70
Richards, Thomas, 234–35, 240
Richardson, Alan, 93, 178
Riemann, Hugo, 164–65
Ritter, Johann Wilhelm, 183
Ritvo, Harriet, 233
Rome, 33
Rossini, Gioachino, 6, 9, 181, 243
Rousseau, Jean-Jacques, 78
Rowley, John, 53
Royal Academy of Music, 13–14
Royal Academy of Painting, 13
Royal Coburg Theatre, 23, 184–85
Royal College of Surgeons, 23, 176–77, 179, 182
Royal Conservatory of Music, 13
Royal Cyclorama, 6
Royal Institution, 5, 13, 115–16, 119–21, 132, 153, 156
Royal Opera Arcade, 11, 102, 145
Royal Panopticon of Science and Art, 17–18
Royal Philharmonic Society, 13–15
Royal Polytechnic Institution, 12, 18
Royal Society, 5, 12, 36, 54, 93, 165, 204, 207
Royal Surrey Zoological Gardens, 9–10
Rubini, Giovanni Battista, 9, 159, 162
Russell Institution, 12
Russell Square, 10

Sacred Harmonic Society, 17
St. Paul's Cathedral, 239
Sampieri, Nicola, 4, 21, 79–97
Sans Pareil Theatre, 104
Sax, Antoine-Joseph (Adolphe), 238
scales and tuning systems, 5, 22, 41, 56, 89, 120, 131–32, 149, 153–56, 160, 162–71
Schaffer, Simon, 36, 42, 71, 103, 209

Scheibler, Johann Heinrich, 131
Scholes, Percy, 238
Schumann, Robert, 78
Sconce, Jeffrey, 137
Scott, John, 104
Scruton, Roger, 77–79, 95–96
Selfridge-Field, Eleanor, 60
Shackleton, Ernest, 163
Shakespeare, William, 18
Shapin, Steve, 36
Shelley, Mary, 23, 175–79, 182–84, 187–88, 194
Shelley, Percy Bysshe, 93, 182–83
Smith, Adam, 188
Smith, O. (actor), 185, 200
Society for the Diffusion of Useful Knowledge, 12
Society of the Concerts of Ancient Music, 14
Soho, 8, 16, 67
South Kensington, 151, 165
South Sea Islands, 165
speaking machine, 5, 22, 116–21, 148–50, 155–62
speaking trumpet, 5, 89
Spectacle Mécanique, 103
Spohr, Louis, 15, 17
Spring Gardens. See Great Room, Spring Gardens
Statistical Society of London, 12
Steele, Richard, 53, 65
Steibelt, Daniel Gottlieb, 86
Stendhal, 181
Stengers, Isabelle, 148
stereograph, 3
Sterne, Jonathan, 36, 132
Stet Sol, 20, 33–34
Strand, the, 16–17, 21, 23, 26, 48, 101, 104–5, 120, 155, 175
Strand Palace Hotel, the, 26
Strauss, Richard, 78–79
street music, 19, 23, 205–20
South Africa, 163, 165, 170
South American Mission Society, 163
Sydenham, 10, 231
symphonium, 115, 118–20, 113

talking machine. See speaking machine
Tallis, John, 236–38
taxonomy, 20, 32, 40, 232–37
telegraphs and telegraphy, 2–3, 16, 22, 101, 104–5, 116–17, 122, 128–29, 137, 140–42, 145–48, 151, 161
temperament. See scales and tuning systems

thaumatrope, 3
Theatre Royal, Birmingham, 65
Theatre Royal, Covent Garden, 15, 66, 184, 186, 196
Theatre Royal, Drury Lane, 10
Theatrical Examiner, 189, 194
Thomas, Sophie, 93, 95
Thompson, Thomas Perronet, 159–60, 168–70
Tolley, Thomas, 64
tongues (vibrating), 114, 156–59, 161–63, 173
Tonic Sol-Fa College, 165, 170
Tramezzani, Diomiro, 192
transparent orrery. *See* eidophusikon
transparent paintings, 14, 20–21, 54, 84
Trevelyan, Charles, 160
trombone, 193, 202
Trower, Shelley, 139, 178
trumpet, 27–28, 80, 135–39
tuning. *See* scales and tuning systems
tuning fork, 22, 111–14, 131–32, 159
Turner, Edward, 8
Tussaud. *See* Madame Tussaud's
typewriter, 3, 145, *146*

University College, 8
University of London, 8

Vasari, Giorgio, 20, 40
Venn, Henry, 160
Verdi, Giuseppe, 9
Victoria, Queen of England, 11, 227, 241
Villa Simonetta, 27–30
violin, 9, 11, 19, 22, 24, 38, 41, 57, 59, 80, 89–90, 116, 131, 181, 202, 218, 237, 242
violoncello, 11, 57, 202
Viotti, Giovanni Battista, 9
vitalism, 23, 176–83, 194–95
Vivaldi, Antonio, 82

Wagner, Richard, 9, 167
Walker, Adam, 4, 7, 20–21, 47–72, *49*, 104

Walker, Dean, *48*, 66, 69–70
Walker, William, 66, 69–70
Watt, James, 18
Weber, Carl Maria von, 9; *Der Freischütz*, 104
Weber, Wilhelm Eduard, 117, 121
Weekly Entertainer, 125
Weeks, Thomas, 103–4
Wesleyan Missionary Society, 16
Westmeath, George Nugent, Marquess of, 211, 217–19
Westminster Abbey, 167
Westminster Mechanics' Institute, 17
Westminster Review, 13, 159–60
Wheatstone, Beata, 101
Wheatstone, Charles, 4–5, 7, 16, 21–22, 101–74;
    Charley Wheatstone's Clever Tricks, 5,
    21, 102, 106; concertina, 3, 22, 118, *119*, *120*,
    145–50, *149*, 161–71; invisible girl, 22, 102,
    135–39; enchanted lyre, 22, 102, 116, 125–
    42, *127–28*, 145–46; Grand Central Dia-
    phonic Orchestra, 102; Musical Museum,
    22, 102–3, 105–6, 121, 125–6; speaking ma-
    chine, 116–18, 150, 156–62; stereoscope, 101;
    symphonium, 118–20, 133; telegraph trans-
    mitter, *147*; telegraphy, 101, 116–17, 128–29,
    137, 140–42, 145–51; Wheatstone bridge,
    101
Wheatstone, William (the elder), 101
Wheatstone, William (the younger), 101
Wheatstone Collection, King's College Lon-
    don, 146, 162
Willis, Richard, 118, 121, 156, 159–61. *See also*
    phthongometer
Wordsworth, 93, 95
Wright, Joseph, 35–36, 72–73

Yorkshire, 51, 227–28
Young, Thomas, 4–5, 21, 89–97, 108, 111, 131

Zoological Society of London, 152